ECOLOGICAL RECLAMATION

in Canada at Century's Turn

Proceedings of the 35th Annual Meeting
of the Canadian Society of Environmental Biologists

ECOLOGICAL RECLAMATION

in Canada at Century's Turn

Proceedings of the 35th Annual Meeting

of the Canadian Society of Environmental Biologists

Hotel Saskatchewan

Regina, Saskatchewan, Canada

September 26-29, 1995

Edited by

Henry T. Epp

Canadian Plains Research Center
University of Regina
Regina, Saskatchewan S4S 0A2
Canada

Canadian Cataloguing in Publication Data

Canadian Society of Environmental Biologists.
Meeting (35th : 1995 : Regina, Sask.)

Ecological reclamation in Canada at century's turn

(Canadian plains proceedings, 0317-6401 ; 28)

Includes bibliographical references.
ISBN 0-88977-102-2

1. Restoration ecology - Prairie Provinces - Congresses.
2. Restoration ecology - Canada - Congresses.
I. Epp, H.T. II. University of Regina. Canadian Plains Research Center. III. Title. IV. Series.

QH541.15.R45C35 1995 333.73'15309712 C96-920170-2

Cover Design: Donna Achtzehner/Agnes Bray

Printed and bound in Canada by
Hignell Printing Limited, Winnipeg, Manitoba

Printed on acid-free paper.

Contents

Foreword

The Canadian Society of Environmental Biologists (CSEB) is a national non-profit organization made up of biologists from across the nation and from other countries as well. Its major objectives are to promote ecologically sound resource and environmental management and to disseminate new, scientifically sound information in this regard. An important means of meeting these objectives is to host annual meetings at various locations across Canada, and to publish the proceedings.

The focal theme of the 35th Annual Meeting of the CSEB, held in Regina, Saskatchewan on September 26-29, 1995, was ecologically based reclamation and restoration of lands and waters disturbed by human activities. Topics included restoration of mined lands, disturbed wetlands, soils disturbed by seeding and irrigation practices, lands and waters disturbed by removal and dumping of urban snow, and more general topics including effects of climate change over a major river basin, and the role of environmental effects (impact) assessment in ecological restoration. Of the papers presented either orally or as posters at the 1995 Annual Meeting, twelve are included here in their edited versions. The final paper is an evaluation of the ecological restoration situation in Canada, and was added after the conference.

At this time, I wish to extend my thanks to Betty Collins, Saskatchewan Water Corporation, for helping me chair the 1995 Annual Meeting, and for providing leadership on the Wednesday (September 27, 1995) field trip to reclamation sites at Estavan, Saskatchewan. Clearly, much of the success of the Annual Meeting and conference is due to Betty's substantial and competent efforts.

Henry T. Epp
Editor

Acknowledgments

The success of the September 26-29, 1995 CSEB conference in Regina was in no small measure due to the inputs of many people. In particular, we want to thank Gail Anderson and Brent Bitter, Saskatchewan Environment and Resource Management (SERM), who did most of the work on conference organization, venue, and information dissemination.

Members of the 1995 CSEB Conference Steering Committee also deserve thanks for their assistance in developing the organization of the conference. In this regard, thanks are extended to Bob Cunningham, Saskatchewan Mining Association; Bernd Martens, Prairie Coal Ltd.; Gerry Fuller, Faculty of Engineering, University of Regina; Malcolm McKee, TAEM Consultants, Ltd.; Stan Saylor, SaskPower; and Maynard Chen, Tom Gates, Malcolm Ross, Brent Bitter and Gail Anderson, SERM.

Several people provided their assistance to make the field trip to Estevan both an enjoyable and a learning experience for the participants. These were Reynold Belitski, SaskWater; Karen Schmeichel, Bruce Duncan, Dwayne Chipley, Mike Pavo and Shelley Crouse, SaskPower; Doug Barnstable, Estevan Coal Corporation; and Gerry Demchuk and Trent Enzsol, Prairie Coal Ltd. We thank Robert Stedwill, SaskPower, for his assistance in providing SaskPower's generous financial and staff support for the field trip.

Special thanks from the CSEB are extended to Les Cooke, Associate Deputy Minister, SERM and to SERM, for providing both financial and moral support to the Annual Meeting and conference and to SERM staff involved in planning and carrying out the work of the conference, and for providing the wrap-up comments at the conference. Without this support this conference would not have been possible.

Lastly, on behalf of the Steering Committee and the entire CSEB membership, we wish to thank Bill Fuller for providing his inspiring perspective on the CSEB and environmental conservation as the Keynote Speaker at the Thursday evening (28 September 1995) banquet.

Henry T. Epp and Betty Collins
Chairpersons, 1995 CSEB Annual Meeting

Prediction to Landscape Restoration: Completing the Environmental Effects Assessment Cycle

by Henry T. Epp

14 Shawbrooke Court SW
Calgary, Alberta T2Y 3G2

Abstract. Environmental impact or effects assessment (EIA, EEA) is a predictive exercise, legislated to proceed via a variety of defined administrative processes in different jurisdictions in the world. An important weakness is the inability to test effects predictions before they occur, and another weakness is infrequent follow-up monitoring providing information useful for post-development cumulative effects assessment and restoration or reclamation of lands and waters. Ideally, EEA is part of a comprehensive regional environmental planning and management process, falling between long-term regional planning and post-development protective regulation and restoration. Ecologically sound EEA and restoration within this greater context are best effected via reestablishing and maintaining landscape integrity. This, in turn, is managed best by ensuring reestablishment and maintenance of biodiversity and ecological processes in relation to the ecosystem components most valued by relevant human groups. The administrative process by means of which this environmental management goal may be effected must be flexible and adaptive so as to incorporate natural environmental change and human cultural dynamics.

Introduction

Environmental impact assessment (EIA), now commonly called environmental effects assessment (EEA), basically is a predictive exercise, legislated to proceed via a variety of defined administrative processes in different jurisdictions around the world. Most processes are regulated by their respective governments, but the actual effects assessments are done by proponents, whether private or public sector. An important weakness in EEA is its inability to test effects predictions before the effects occur (Beanlands and Duinker 1983; Epp 1995b). This is because the decision on whether to approve a proposed development has to be made by the regulatory authority before the development proceeds and, hence, before the effects can occur. An implication to the science involved in EEA creates a further weakness in the process. While EEA may use scientific information to make predictions, it cannot carry deductive science to its logical conclusion, predictions testing, before the decision on whether to approve the proposed development has to be made by the respective jurisdiction.

Due to its project orientation, EEA forms only part of the environmental protection process in any environmentally responsible jurisdiction. Various forms of planning processes and protective legislation make up the balance. EEA essentially is an important front-end process to project planning. Ideally, EEA lies between long-term regional planning and standards or other protective regulation. In addition, EEA always deals with prospective negative environmental effects which may lead to future problems. Hence, the logical end to the EEA process should be ecologically sound reclamation, restoration of ecological processes in the potentially affected area to the nearest possible approximation of the original natural or baseline condition.

Predictive-deductive methodology is an important requisite to reductionist science which, in turn, provides the verification required to create credible building blocks for synthesis, holistic or "big picture" science. Knowledge of holistic ecological processes is essential to understanding environmental health, and without this understanding efforts at maintaining environmental health may be wasted. These processes require a high level of biodiversity in any defined area, and this biodiversity is best retained by maintaining landscape integrity (Epp 1992). Hence, adjustments to environmental effects assessment such as follow-up testing of predictions, with subsequent input into regional planning for landscape integrity, should provide a sound instrument for holistic environmental management, including coping with cumulative effects. *The objective of this paper is to develop a structure for integrating science and environmental effects assessment and follow-up, providing a basis for establishing post-development reclamation or restoration plans with the goal of maintaining long-term landscape ecosystem integrity.*

EEA as Part of Scientific Environmental Management

Environmental effects assessment, simply put, is predicting the difference between the baseline environment of a prospective development and the new environment following changes caused by the development. "An environmental impact, which is what the EEA predicts and evaluates, is the difference between a baseline measurement and a predicted state of a defined environmental entity once the project is underway" (Epp 1995a, partly after Duinker 1989). The universal dilemma of EEA as an isolated process is that the product, the environmental impact statement (EIS), is based on prediction, not measurement and testing; hence, it is never possible to know what the precise effects will be before the decision on whether to proceed with the development must be made (Duinker 1989). Consequently, it follows that EEA is best used as part of a more comprehensive scientifically based environmental management process, between project planning and mitigation/follow-up monitoring.

The step between EEA approval and what happens after is often not part of the EEA process, as mentioned previously. The result is a communications gap between EEA and regulatory administrative processes. EEA follow-up, leading back to testing the predictions for accuracy, and cumulative effects assessment (CEA) work, then, when dealt with at all, most often becomes the responsibility of regulatory agencies only, rather than being factored into the EEA process to feed back into future assessments. Clearly, there is need for communication between EEA and regulatory agencies to eliminate this gap, as the need for predictions accuracy and CEA becomes ever more important in areas of rapidly expanding development. A sound restoration plan, then, is dependent upon EEA follow-up predictions testing to provide accurate information, and restoration follow-up testing is requisite to technically sound implementation of the plan over the long term. Thus, follow-up testing and information flow form the structure and process for completing the EEA-restoration cycle.

All of this planning, prediction, assessment, and testing must be translated into action on the ground to achieve ecologically sound restoration following development disturbance. Landscape restoration has two fundamental optional goals, based on the recognition that absolute restoration to the precise original environment never is possible: (1) return of the landscape to as near to its original condition as possible and maintaining it as such, its integrity, and (2) formation of a new landscape. Clearly, from an environmental viewpoint, the first option is preferable.

Fundamental to understanding the requirements of meeting the goal of restoring and maintaining landscape integrity is the concept of landscape itself as applied to ecology, namely, landscape ecology. The idea of landscape is not new, likely having originated in Holland as an expression of human cultural influences on the land in the form of pictures, "landscape" paintings (Zonneveld 1988). This concept evolved into the dual natural and cultural Dutch "landschap" and German "Landschaft" or comprehensive landscape concept, a landscape being an "area made up of a distinct association of forms, both physical and cultural" (Sauer 1925). Hence the terms "natural landscape" and "cultural landscape" (cf. Forman and Godron 1986; Berglund 1991).

Application of the landscape concept to land-based disciplines has resulted in the term "landscape ecology" (Zonneveld 1988). Beginning first in Germany following the efforts of Troll (1950), and finally becoming established in North America in the 1980s, landscape ecology has become a recognized and rapidly expanding subdiscipline at the forefront of developing the environmental conservationist paradigm, a "shift in attention to the world's landscape homes" (Rowe 1988). In applying the landscape ecology concept to conservation and reclamation techniques, it is necessary to view the "landscape ecosystem" as composed of three subsystems, the "geosystem, biosystem and anthroposystem" (Leser and Rodd 1991), combining the natural and cultural landscapes to meet conservation and environmental protection goals.

The landscape concept is "elusive" in a cognitive sense because a landscape is whatever the definer wishes it to be (Forman and Godron 1986). Yet this very subjectivity is what makes landscape a powerful conceptual tool because it transcends empirical science into human-centred resource strategy and decision making (Naveh and Lieberman 1984). In fact, the landscape as ecosystem paradigm springs from the melding of two disciplines, ecology and geography, both of which rely on definition of physical parameters for overtly stated purpose (Rowe and Sheard 1981). Methodological strength lies in the definition of the premise, maintaining scientific rigour by retaining deductive potential. Hence, landscape ecology is at once (1) a useful scientific explanatory tool, (2) has application to real life conservation, and (3) has the appeal providing the public acceptability for post-decision making implementation of conservation plans or strategies. The idea of landscape, then, allows the concept of the ecosystem to be expressed in operational terms, including application to environmental improvement of disturbed lands, the landscape ecosystem.

Given this paper's overall restoration/reclamation goal of maintaining long-term landscape ecosystem integrity, what, then, constitutes such integrity? Accepting the definition of landscape ecosystem as composed of the geosystem, the biosystem, and the anthroposystem, it follows that the means to landscape ecosystem integrity are to return altered lands and waters as nearly as possible to their extant (1) biodiversity, (2) landforms, (3) drainage patterns, (4) ecological processes, and (5) valued (by humans) ecosystem components (VECs).

As explained previously (Epp 1995a, b), the application of science to EEA normally is restricted to ensuring accurate technical and environmental information, making inductively derived generalizations from the data, and deducing the needed predictions from these generalizations. Unfortunately, the EEA process normally ends there, not permitting testing of the predictions for their falsity or truth, as the decision on whether to proceed with a development has to be made prior to the occurring of the environmental effects. Hence, a full application of science to EEA requires follow-up monitoring to provide data for the test to determine the accuracy of the predictions. In a situation in which restoration/reclamation is a condition of the EEA approval to proceed with the development, the actual long-term effects are post-restoration. This requires post-restoration testing for confirmation of the predictions to establish scientific validity of the results.

Post-EEA Restoration Planning

Having defined the landscape ecosystem concept, and having indicated how science may strengthen EEA and follow-up, it remains to develop a means to integrate this process into developing sound restoration plans. When implemented, these plans will help maintain long-term landscape ecosystem integrity. This integrity, in turn, requires maintenance of nearly extant levels of biodiversity, ecological processes, landforms, watershed features, and valued ecosystem components.

Biodiversity and ecosystem processes require physical connecting links among living populations to prevent island-like isolation which leads to loss of diversity and extinctions, simplification of ecological processes, and, ultimately, to more extinctions (MacArthur and Wilson 1967). This is a positive feedback situation dangerous to the integrity of any living system (Berryman and Millstein 1989). In large natural wildernesses these connections rarely are severed due to the effects of large scale, where any chance disturbance in one place is unlikely to be universal so that loss of biodiversity and processes is rare. In environments heavily committed to human use, however, separation of populations is common. The only way to maintain genetic connection is via natural or consciously designed conduits along which genetic information can travel freely in both directions (Epp 1995c). In heavily developed environments, each further development adds cumulatively to limiting the connections. Hence, any ecologically sound restoration plan must be designed to retain enough landscape integrity so as to maintain sufficient genetic conduits, or else it will fail in the goals of maintaining biodiversity and ecological processes.

Restoration is after development mitigation. The EEA process, applied before the development proceeds, should set the stage for a combination of preventative and mitigative (restoration or reclamation) activities. The process diagram shown in Figure 1 illustrates how EEA and restoration activities may be integrated to attain maximum environmental protection via reestablishing landscape integrity.

Much restoration is highly specialized technical work, demanding precise knowledge about baseline soil structure and chemistry, hydrologic conditions, ecological processes, and human uses and values, the pre-development landscape. This information, then, is applied to the post-development conditions so as to return the affected environment as near to the baseline condition as possible. Sound baseline data obtained during the EEA phase of the planning-development-restoration cycle save time and effort later when the need is urgent to restore the affected environment.

Ensuring original landscape integrity begins with a return of disturbed lands to their physical baseline. Appropriate equipment applied in the correct places with saving and redeposition of topsoil usually suffices. Return to the original biodiversity and ecological processes is more difficult to attain, however, because these conditions are much more subjectively and often less well defined, partly because natural process change occurs continually, even during the time of disturbance. Hence, while an important restoration goal may be to retain and maintain ecological processes, the simplest and most effective approach to attaining this end is to reestablish and maintain the physical and biological conditions that are requisite to the continued functioning of these processes, albeit changing processes. This means ensuring re-creation and retention of the habitats necessary for the original biodiversity.

Restoring and protecting the geosystem, although it may be costly, is easier than doing the same for the biosystem and anthroposystem. Information tends to be more measurable and precise for physical systems. Essentially, this process requires ensuring retention of the hydrologic system, both surface and ground waters, via replacement of the original contours, topsoil, and surface materials so as not to alter seepage and ground water flows significantly.

Reestablishing the original biodiversity via habitat restoration requires attention to the flora first, ensuring that native plant species are reestablished to a reasonable approximation of the baseline condition. Native animals then may recolonize the restored landscape on their own, but careful monitoring is needed so that should they fail to do so, artificial recolonization may proceed. Equally important, monitoring is needed to ensure that continuity with nearby native populations is not lost and, if it is lost, that quick action will take place to prevent local extinction, otherwise the expensive reestablishment process must begin over again or there will be serious loss of biodiversity.

Why the attention to retention of biodiversity within the biosystem? The most recent theoretical advancements in ecological theory have the ecosystem behaving as a complex self-organizing system (cf. Hollick 1993). A self-organizing system has a pre-existing structure through which energy flows, and which reacts to disturbance by feeding products back into the system which may increase production (Odum 1988). Alternatively, system feedback also can cause a downward spiral, decreasing production or greatly increasing unwanted products, possibly leading to chaotic oscillation that leads inexorably to unpredictability of outcome, an undesirable situation from an environmental management viewpoint. Clearly, then, biodiversity in an ecosystem is

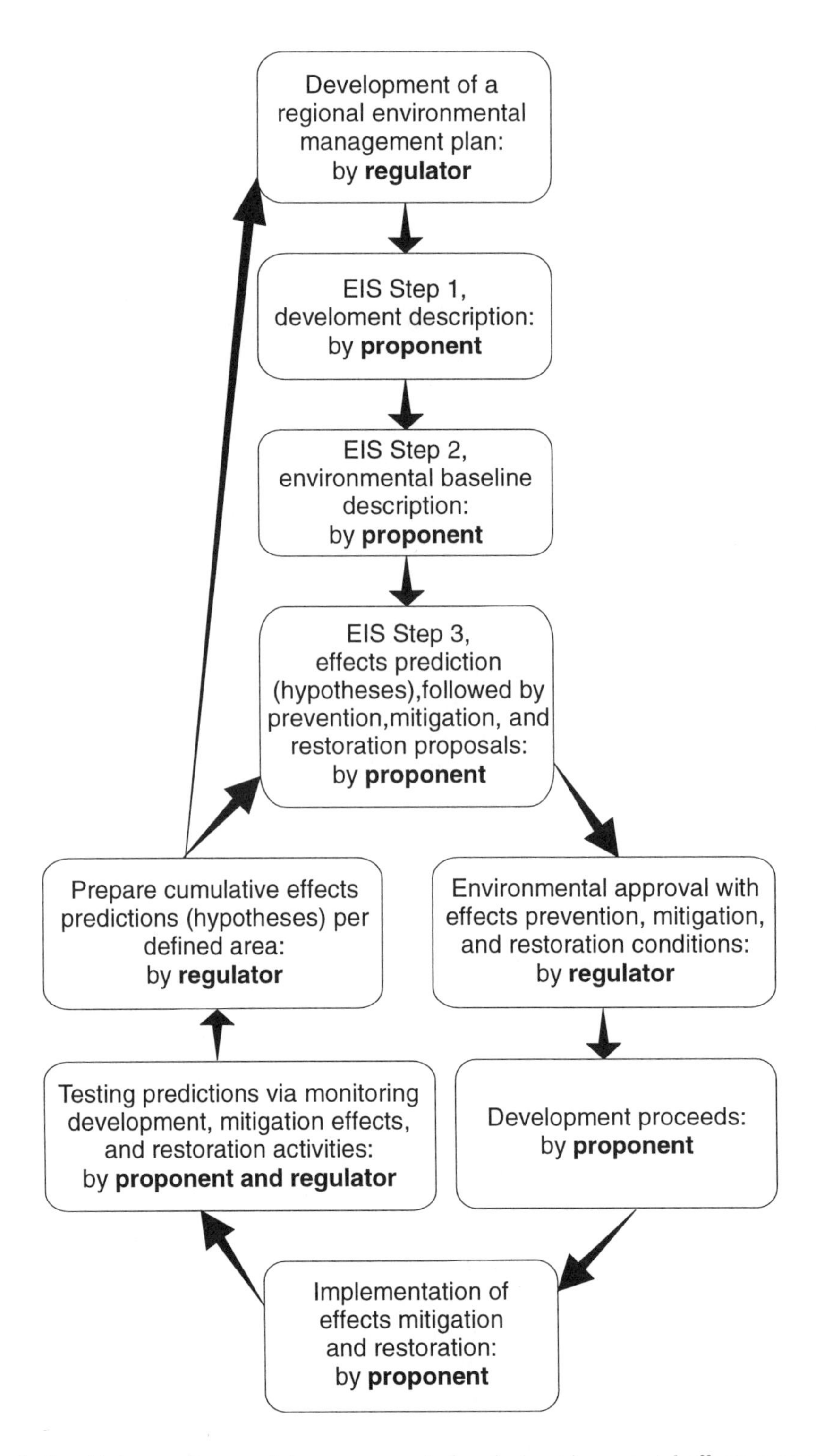

Figure 1. Combining environmental management planning, environmental effects assessment, and restoration into one cyclical regulatory and administrative process.

linked positively to resilience to outside disturbance. A normally functioning, diverse, and resilient ecosystem, then, is much superior from a long-term management view than is an artificially maintained ecosystem no matter how productive or attractive it may seem. Hence, reestablishing and maintaining biodiversity is priority number one in any ecologically sound reclamation plan.

The anthroposystem or human factor in landscape restoration relates mostly to valued ecosystem components. These VECs vary with the human culture affected and also with subcultures and geographic locations, as lifestyles may be partially dependent on the local environments. For example, in the agrarian Canadian prairies productive grain lands are valued highly, in ranching country productive grasslands are important, and in urban centres parks and other wild lands and waters, often far distant from a city, may be highly valued. In the boreal forest traditional uses remain important in many locations, so that the resources required for the traditional lifestyle, such as moose, caribou, furbearers, and fish, are the primary VECs.

Conclusions

Theory and practice always are inextricably intertwined in any human endeavour, environmental management, environmental effects assessment, and restoration or reclamation included. The theory always drives the practice, but information and experience gained from practice ideally should feed back into theory. Continued ecologically sound restoration, then, rests on advancements in ecosystem theory as much as it does on technological advances, and it is as subject to change as the ecosystem itself.

The objective of this paper has been to integrate post-development restoration into environmental effects assessment and, ultimately, into comprehensive environmental planning and management. The theoretical underpinnings of this proposed exercise ideally relate to reestablishing and maintaining the integrity of the affected landscape ecosystem.

Emphasis in this paper is placed on the biosystem, as this is the most difficult part of the ecosystem to reestablish and maintain. Complex self-organizing systems theory applies here, as it is the ecosystem processes that must be reestablished and preserved, not necessarily any individual physical feature or biological population in its original defined baseline state. This implementation process is maintained most effectively by retention of biodiversity, the living requisite for ecosystem resilience in the face of external disturbances such as human developments preceding restoration. A hypothetical administrative flow structure is provided in Figure 1 to indicate how the objective of this paper may be met in practice via integrating environmental planning, environmental effects assessment, and restoration into one comprehensive environmental management process.

Ecological restoration applies especially to cumulative effects of developments within a defined landscape over time, as the ecosystem cannot be managed for landscape integrity on an individual project-by-project basis. Biodiversity and ecological, physical, and cultural processes involve large areas and are diachronic, not circumscribed by time. Furthermore, they are changing constantly, so that ecologically and culturally sound restoration becomes part of adaptive management in Holling's

(1978) sense. Hence, maintaining these landscape processes via ensuring biodiversity and landscape integrity requires feeding information from post-environmental effects assessment monitoring back into environmental planning and EEA. This is a changing, flexible, and adaptive environmental planning and management process that incorporates natural and cultural changes into a dynamic administrative process, meeting the objective of this paper.

Acknowledgments

I thank Brent Bitter for computerizing the diagram presented in this paper. The opinions expressed in this paper are my own, and do not necessarily reflect or represent those of anyone else.

References

Beanlands, G.E., and P.N. Duinker. 1983. *An ecological framework for environmental impact assessment in Canada.* Environment Canada, Hull, Quebec .

Berglund, B.E., editor. 1991. The cultural landscape during 6000 years in southern Sweden - the Ystad project. *Ecological Bulletins* 41. Munksgaard International Booksellers and Publishers, Copenhagen.

Berryman, A.A., and J.A, Millstein. 1989. Are ecological systems chaotic? *Trends in Ecology and Evolution* 4 :26-28.

Duinker, P.N. 1989. Ecological effects monitoring and impact assessment: what can it accomplish? *Environmental Management* 13:797-805.

Epp, H.T. 1992. Facing the ecosystem conundrum: ecological and economic productivity-stability divergence. In *Landscape approaches to wildlife and ecosystem management:*19-26. Edited by G.B. Ingram and M.R. Moss. Proceedings of the Second Symposium of the Canadian Society for Landscape Ecology and Management, University of British Columbia, Vancouver, June 1990. Polyscience Publications Inc., Morin Heights, Quebec.

Epp, H.T. 1995a. Role of scientific prediction in environmental impact assessment. In *Environmental impact assessment and remediation: towards 2000:* 91-98. Edited by W.T. Dushenko, H.E. Poll, and K. Johnston. Proceedings of the 34th Annual Meeting of the Canadian Society of Environmental Biologists, Victoria, British Columbia, June 1-4, 1994. Canadian Society of Environmental Biologists, Toronto, Ontario.

Epp, H.T. 1995b. Application of science to environmental impact assessment in boreal forest management: the Saskatchewan example. *Water, Air and Soil Pollution* 74:179-188.

Epp, H.T. 1995c. Churchill River heritage: role in biodiversity. In *The Churchill: a Canadian heritage river:*18-29. Edited by Peter Jonker. Extension Division, University of Saskatchewan, Saskatoon.

Forman, R.T.T., and M. Godron. 1986. *Landscape ecology.* John Wiley & Sons, New York.

Hollick, M. 1993. Self-organizing systems and environmental management. *Environmental Management* 17:621-628.

Holling, C.S., editor. 1978. *Adaptive environmental assessment and management.* John Wiley & Sons, New York.

Leser, H., and H. Rodd. 1991. Landscape ecology — fundamentals, aims and perspectives. In *Modern ecology: basic and applied concepts:* 831-844. Edited by G. Esser and D. Overdiek. Elsevier, Amsterdam.

MacArthur, R.H., and E.O. Wilson. 1967. *The theory of island biogeography.* Princeton University Press, Princeton, New Jersey.

Naveh, Z., and S. Lieberman. 1984. *Landscape ecology - theory and application.* Springer, Berlin.

Rowe, J.S. 1988. Landscape ecology: the study of terrain ecosystems. In *Landscape ecology and management:* 35-42. Edited by W.R. Moss. Proceedings of the First Symposium of the Canadian Society for Landscape Ecology and Management, University of Guelph, May, 1987. Polyscience Publications Inc., Montreal, Quebec.

Rowe, J.S., and J.W. Sheard. 1981. Ecological land classification: a survey approach. *Environmental Management* 5:451-464.

Sauer, C.O. 1925. The morphology of landscape. *University of California Publications in Geography* 2(2):19-24.

Zonneveld, I.S. 1988. Landscape ecology and its application. In *Landscape ecology and management:* 3-15. Edited by M.R. Moss. Proceedings of the First Symposium of the Canadian Society for Landscape Ecology and Management, University of Guelph, May 1987. Polyscience Publications Inc., Morin Heights, Quebec.

The Montreal, Quebec Experience in Used Snow Disposal and Treatment

by Claude E. Delisle and Pierre André

Groupe Neige / Snow Research Group
École Polytechnique de Montréal, P.O. Box 6079, station Centre-ville, Montreal, Quebec, Canada, H3C 3A7 and Département de géographie, Université de Montréal, P.O. Box 6128, station Centre-ville, Montreal, Quebec, Canada H3C 3J7

Abstract. In 1990, the Association des Biologistes du Québec organized a symposium with the Saint Lawrence Centre (Environment Canada) entitled: "The Saint-Lawrence: a river to be reclaimed" (ABQ 1990). A paper on used snow then was presented explaining how snow management practices can be modified to prevent contaminant input into the Montreal harbour waters. Taking used snow as an example, the present paper exposes how far we should go in the reclaiming procedure. In urban and industrial areas a zero discharge objective remains excessive for used snow and is economically hard to achieve. An average of 250 cm of snow falls on the city of Montreal every winter. There are from eight to ten major snowstorms each season, requiring snow removal along 1900 km of streets and 3200 km of sidewalks. This operation annually produces a volume of about 7,000,000 m^3 of used snow which is disposed through five methods: quarry dumping, snow chutes into sewers, surface sites, snow melting, and river dumping. The latter method will not be allowed beyond 1996 due to regulation from the Quebec Ministry of Environment. Of the 30,000,000 m^3 used snow in the Province of Quebec more than 30% is discharged directly into lakes and rivers (André and Delisle 1989). Once it has fallen to the ground, snow stays on the streets for a short period (24 to 96 hours) before it is plowed and cleared away. During this time, the snow accumulates de-icers (e.g. NaCl) and urban pollutants of all kinds (e.g. Pb, Cr) and this whole mixture is designated as "used snow." Since 1984, the city of Montreal and suburbs have carried out environmental studies to learn more about the quality of used snow in an urban area and its potential threat to the environment. Research on used snow contamination has been done along two kinds of streets. Commercial and residential arteries were selected and results showed that significant differences existed between these sites. Equally different results were found between residential and commercial sectors when the used snow was taken from transport trucks. The duration that used snow stays piled up along the curbside also influenced its concentration of contaminants. The average concentrations of contaminants in used snow after truck loading were high in suspended solids, chlorides, sulphate, oil and grease, calcium and sodium, and also were high in some heavy metals (Fe, Pb). This is explained by the fact that curbside used snow is plowed along the street and accumulates contaminants through contact with the road pollutants during loading operations. The Quebec Ministry of Environment and Fauna policy regarding used snow will be discussed (MENVIQ 1988) and compared with urban runoff problem and sewer overflow, which are of equivalent importance.

Introduction

An average of 250 cm of snow falls on the city of Montreal's territory every winter. There are from eight to ten major snowstorms each season, requiring snow removal along 1900 km of streets and 3200 km of sidewalks. This operation annually produces a volume of about 7,000,000 cubic metres (m^3) of used snow disposed of through five methods (quarry dumping, snow chutes on sewers, surface sites, snow melting, and river dumping; the latter will not be in use by fall 1996 due to regulation from the Quebec Ministry of Environment, Table 1). More than 30,000,000 m^3 of used snow must be cleaned up in the Province of Quebec, Canada, out of which 30% now is directly discharged into lakes and rivers (André and Delisle 1989; Delisle 1993, 1995).

Once it has fallen to the ground, snow stays on the streets for a short period (24 to 96 hours) before it is plowed and cleared away. During this time, the snow accumulates de-icers (e.g. NaCl), abrasives and urban pollutants of all kinds (e.g. Pb, Cr), and this whole mixture is designated as "used snow." Since 1984, the city of Montreal and suburbs have carried out environmental studies to learn more about the quality of used snow in an urban area and its potential threat to the environment.

Research on used snow contamination has been done along two kinds of streets. Commercial and residential arteries were selected and results showed that significant differences existed between these sites (Table 3).

Equally divergent results were found between residential and commercial sectors when the used snow was taken from transport trucks. The lapse of time that used snow stays piled up along the curb also influences its concentration of contaminants and their potential ecotoxicological effects when dumped into bodies of water. Determination of an ecotoxicological potential related to used snow during winter 1996 should give more information on its toxicity.

Average concentrations of contaminants in used snow after loading are high in suspended solids, chlorides, sulphate, oils and greases, calcium and sodium, and also are high in heavy metals (Fe, Pb, Cr). This is explained by the fact that curbside used snow is plowed along the street and becomes contaminated through contact with the road pollutants before and during loading operations.

Used snow contains three major pollutants: lead (Pb), suspended solids (SS), and chlorides (Cl-). Of the heavy metals present in the snow, lead is considered the most severe pollutant and is found in both soluble and insoluble forms (Riverin 1982), its toxicity depending on the chemical status of the pollutant (Landsberger 1984). The soluble form is the most toxic, and it is also very stable and difficult to eliminate (Scott 1980). Manganese [methylcyclopentadienyl manganese tricarbonyl (MMT)], which now replaces lead in gasoline, also should be considered for analysis (Loranger and Zayed 1995).

When analyzing "used snow" for the concentrations of different elements, the corresponding values of the "natural snow" also must be appreciated. Geographic locations, geological conditions, and storm origins may affect the quality of the natural snow (Percherkin and Burmatova 1965). The falling snow also is exposed to a number of air pollutants (Landsberger 1983), which will vary depending on the area, as reported by Lewis et al. (1983), who report on lead emissions in Montreal affecting the air

Tableau 1. Méthodes de disposition des neiges usées en pratique sur le territoire de la CUM et de la province de Québec
(Table 1. Snow disposal methods on the MUC and Province of Québec)

Méthodes	Montréal (MUC) 1990 %	Province de Québec 1993 %	Impacts environnementaux
Déchargement dans les cours d'eau (River dumping)	30	30	++++(1)
Déchargement en carrière (Quarry dumping)	24	±5	++(2)
Déchargement en égouts (collecteur) (Snow chutes into sewers)	25		++(3)
Dépôts de surface (plusieurs types) (Surface sites disposal)	20	65	+++(4)
Fondeuses (Snow melters)	1	0	++(5)
Disposition sur terrains privés (Disposal on private land) (souffleuse/snow blower)	0	De + en + fréquent	+++(6)
Refoulement en bordure des routes (Side road disposal)	0	peu	+(7)

(1) Impacts majeurs surtout avec des neiges usées originant de milieux industriels et commerciaux.
(2) Impacts moyens si les formatoins géologiques sont non fracturées si l'eau de fonte est pompée vers une station d'épuration.
(3) Impacts moyens s'il n'y a pas de débordements du réseau collecteur en temps d'orage et de fonte. Nuisances par le bruit et à la circulation automobile.
(4) Impacts moyens pouvant être atténués si les eaux de fonte sont traités par des bassins de décantation et si les sites sont nettoyés au printemps. Sites devant être éloignés des cours d'eau et les eaux de fonte dirigés vers une station d'épuration.
(5) Impacts moyens si les eaux de fonte sont rejetées aux égouts vers une station d'épuration. Génération de pollution atmosphérique par combustion et les embruns.
(6) Impacts sur la végétation et l'esthétique des lieux. Acceptation sociale difficile.
(7) Impacts mineurs si ce n'est que de l'effet des sels de déglaçage sur la végétation. Bonne alternative le long des routes en milieu rural et faiblement urbanisé.

composition. It is important to be aware of such initial differentiation, both of anthropogenic and natural origins, when measuring pollutants in used snow. The manganese (Mn) contamination also is of importance now in Quebec and in the rest of Canada. Some data on manganese concentrations are found in Table 3.

The present study describes the used snow physico-chemical qualities found in the Montreal area, while concentrating on the content of its various pollutants and the best methods of disposal and treatment (Figures 1 to 3).

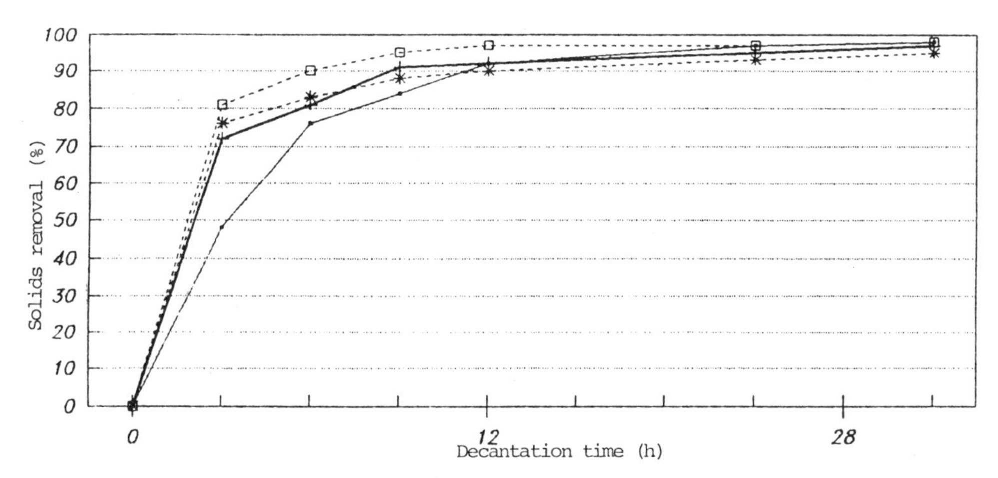

Figure 1. Suspended solids decantation from melted used snow of four different surface dumped sites (sites: Ste-Rose, Polytechnique, Contrecoeur, Bates).

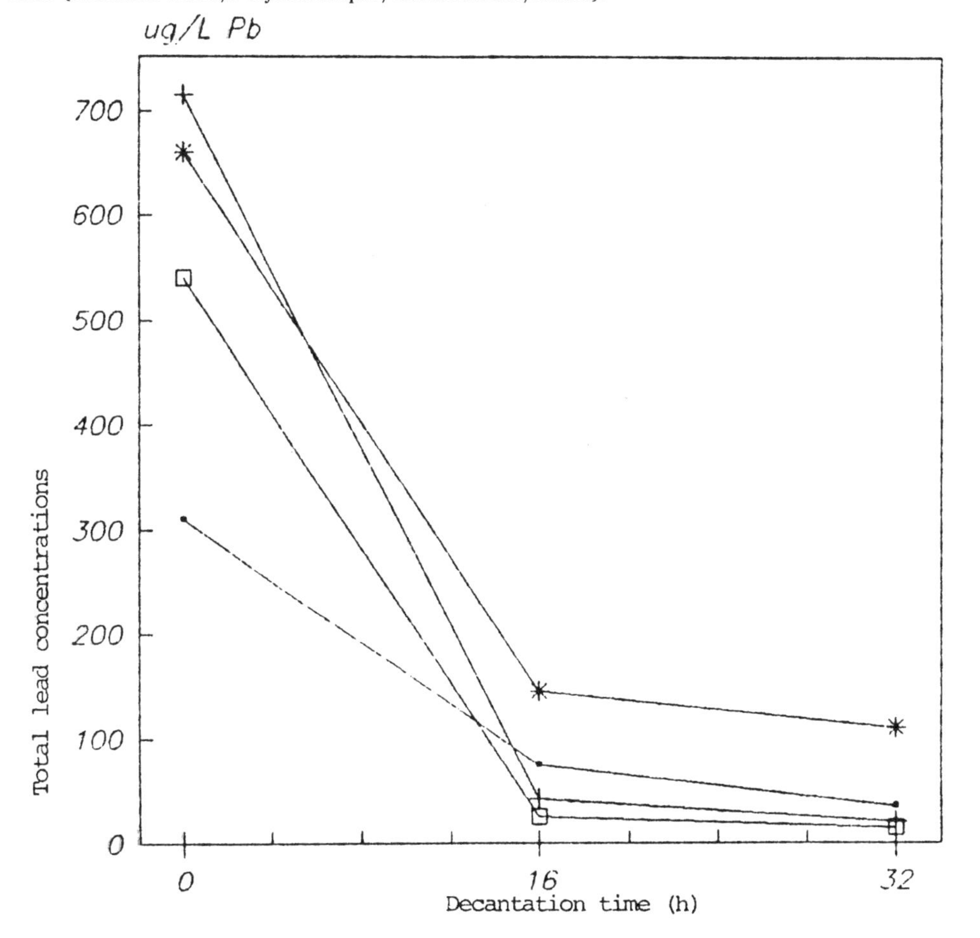

Figure 2. Total lead decantation from melted snow of four different surface dumped sites (sites: Ste-Rose, Polytechnique, Contrecoeur, Bates).

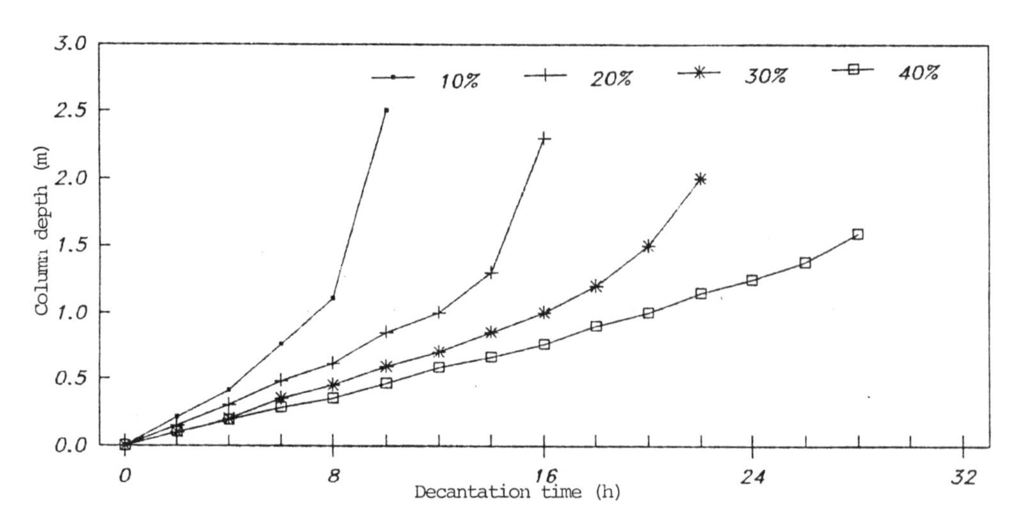

Figure 3. Decantation of particles at the Contrecoeur runoff snow melted from a column test (site: Contrecoeur, city of Montreal).

Materials and Methods

With the exception of temperature, laboratory analyses were carried out 24 hours after collection, allowing the snow to melt in hermetically closed plastic containers, at room temperature. The analyses were done according to the Standard Methods (APHA 1989) and by atomic absorption (AA) for the metals. The number of analyses varied from 12 to 108 (Tables 2 and 3) in relation to the 28 parameters chosen to reflect the literature review on the subject.

Results and Discussion

Used snow global quality. Table 2, taken from Zinger and Delisle (1988), presents a summary of the results obtained from snow sampled in trucks. The great variations in concentrations between the minima and maxima are due mainly to the origin of the snow collected and to the long time it remained in the streets before collection. Contamination of the snow is directly proportional to the residence time in the streets and sidewalks (Leduc and Delisle 1990). Results from Table 3 show lower contaminant concentrations due to fast removal of the snow and their residential origin compared to commercial origin.

Suspended solids and lead. One of the most important contaminants of used snow is suspended solids on which many pollutants are adsorbed. Concentrations up to 8546 mg L^{-1} were found (Table 1). But also true is the fast removal of these solids through decantation (Figure 1). In fact, a short period of four hours removes 60% to 85% of the suspended solids contained in a snowpack taken from four different surface sites in the city of Montreal. These results are independent of the snow origin. Total lead removal is correlated with suspended solids decantation. The lead concentrations also decreased rapidly (± 16 hours) with time (Figure 2).

Table 2. Summary of the physico-chemical characteristic analysis of the used snow

Substances	Units	Number of samples	Mean	Minimum	Maximum	Standard Deviation
pH	-	108	8,5	6,8	9,9	0,39
Temperature	°C	108	-3,03	-8,0	0,0	2,13
Turbidity	UTN	108	29,8	5,0	90,0	19,41
Conductivity	μmHos	108	11128,7	150,0	26500,0	7451,76
Suspended solids	mg L^{-1}	108	1209,0	86,0	8546,0	1186,64
Chlorides	mg L^{-1}	98	3851,2	56,0	10000,0	2335,72
Hardness	mg L^{-1}	98	495,8	23,0	1680,0	380,85
BOD	mg L^{-1}	19	7,5	2,6	13,8	3,36
COD	mg L^{-1}	98	496,6	46,8	1926,8	349,22
Nitrates	mg L^{-1}	93	5,2	1,2	14,9	2,88
Tot. sol.	mg L^{-1}	98	6948,8	1039,0	37359,0	6948,75
Inorganic phosphates	mg L^{-1}	98	3,6	0,12	20,6	3,84
Na	mg L^{-1}	98	4049,6	1000,0	13600,0	2315,02
Ca	mg L^{-1}	96	146,4	34,0	500,0	108,88
K	mg L^{-1}	98	10,6	2,2	45,0	1,54
Mg	mg L^{-1}	98	2,6	0,5	7,0	6,92
Cyanide	mg L^{-1}	32	0,24	0,12	0,33	-
Sulphates	mg L^{-1}	30	129,8	25,0	295,0	65,71
Ammonia nitrogen	mg L^{-1}	30	0,4	0,1	0,8	0,12
Oil and greases	mg L^{-1}	30	104,6	9,0	200,0	56,11
Debris	mg L^{-1}	93	5888,1	542,6	32542,1	6430,53
Pb (total)	mg L^{-1}	93	84,84	17,15	360,05	63,26
Fe (total)	mg L^{-1}	93	912,57	238,6	3762,52	632,00
Cu (total)	mg L^{-1}	93	9,36	1,99	100,48	10,79
Zn (total)	mg L^{-1}	93	42,77	10,02	221,25	33,03
Cr (total)	mg L^{-1}	93	6,67	1,3	35,95	5,89
Cd (total)	mg L^{-1}	30	0,27	0,23	0,40	0,04
Hg (soluble)	μg L^{-1}	12	0,250	0,087	0,137	-

[From Zinger and Delisle 1988. Note: These snow samples were taken in the truck box and show high levels in comparison to our new methodology of sampling a 20 L volume of snow from the snowbank (see Table 3 for post-1988 results)].

Tableau 3. Concentrations des paramètres physico-chimiques retrouvées dans des neiges usées par diverses études
(Table 3. Concentrations of physico-chemical parameters found in used snow after the 1988 study)

Auteurs (authors)	Zinger & Delisle	Delisle & Leduc	Delisle & Leduc	Delisle & Lapointe	Delisle & al.
Année (Year)	1988	1990 Résidentiel	1990 Commercial	1991 Mixte	1993 Résidentiel
Méthode (Sampling)	Camions (Truck)	Carottage (Core)	Carottage (Core)	Tranche (Transect)	Tranche (Transect)
Nombre de Sites	108	12	12	55	299
Conductivité (mS/cm)	11,13	2,01	9,71	5,82	4,90
Écart Type (Standard Deviation)	7,45	2,15	9,46	7,15	4,80
Minimum	0,15	0,10	0,05	0,10	0,05
Maximum	26,50	9,53	41,58	29,03	64,20
Nombre de Sites	108	88	70	609	299
Chlorures (mg/L) (Cl-)	3851,2	508,5	2647,8	2021,0	2072,7
Écart Type (S. D.)	2335,7	649,1	1968,1	2178,0	2190,8
Minimum	56,0	23,0	66,0	33,0	8,6
Maximum	10000,0	7000,0	9500,0	9927,0	29587,0
Nombre de Sites	98	270	209	574	299
M.E.S. (mg/L) (Suspended particules)	1209,0	103,9	355,9	2057,0	496,6
Écart Type (S. D.)	1186,6	101,4	355,9	2057,0	312,3
Minimum	86,0	50,0	16,0	50,0	5,0
Maximum	8546,0	480,0	854,0	16270,0	2318,0
Nombre de Sites	108	270	209	609	299
HUILES & GRAISSES (mg/L) (Oil & greases)	104,6	11,9	20,8	29,0	13,1
Écart Type (S. D.)	56,1	6,7	10,1	31,0	11,1
Minimum	9,0	2,0	1,0	1,0	1,1
Maximum	200,0	28,0	43,0	150,0	45,0
Nombre de Sites	30	47	39	523	33
PLOMB TOTAL (mg/L) (Lead-Pb)	84,84	0,21	0,51	0,65	0,10
Écart Type (S. D.)	63,26	0,29	0,25	0,67	0,09
Minimum	17,15	0,01	0,07	0,01	0,002
Maximum	360,05	2,59	1,40	4,55	0,68
Nombre de Sites	93	270	209	608	299
MANGANÈSE TOTAL (mg/L) Mn	—	—	—	—	0,17
Écart Type (S. D.)	—	—	—	—	0,15
Minimum	—	—	—	—	0,003
Maximum	—	—	—	—	1,08
Nombre de Sites	—	—	—	—	210

From these results, an easy treatment by decantation ponds of the snowpack accumulated at different surface sites seems to be desirable. Unfortunately, decantation trials are lacking (Pinard et al. 1989), and the column tests done with runoff snow melted water at the Contrecoeur, Montreal surface site containing 2000 mg/L of fine particles (2 μm) show the weak in-situ efficiency of such a treatment (Figure 3). During spring snowmelt of the surface site the runoff decantation time is a function of the retention time in the pond. This cannot exceed 8 to 16 hours due to high water flow and the available surface land in an urban area. A retention of only 20% is possible after 16 hours of decantation time. However, larger particles will show a greater decantation in the treatment pond.

Interaction to solve the problem? The dilemma of the road department decision-maker can be difficult. Some citizens request "wall to wall" service level and policies and are not ready to modify or slow down their way of life during winter. Others may accept, for example, snow dumping on their property front, the use of less de-icing salt on their street, etc. It is important that an *interaction* happen between decision makers, citizens, environmental groups and the Ministry of Environment to decide on the service level necessary for snow removal and treatment. A budget of more than 50 millions dollars is needed yearly just for Montreal snow management. Is there a way to cut without damaging the environment and still offering security to citizens?

Conclusion

When studying used snow disposal, treatment, and ecological effects, it is important to identify the real pollution problem associated with all the water pollution sources. Priorities should be given to industrial effluent, discharged, with or without treatment, on a daily basis, day and night, 365 days/year. Urban runoff and storm sewer runoff also are more important than used snow as a pollution source. In Montreal, runoff is active more than 50 days/year, whereas used snow dumping occurs only 20-25 days/year with a much reduced volume ($\approx$ 2,000,000 m^3).

From our snow characterization results, we believe that if a snowstorm of more than 10-15 cm is removed rapidly (24 to 48 hours after the storm) on *residential* streets where no salt and abrasive have been applied, this snow can be disposed of into a body of water without any noticeable ecological effect. Our 1996 study on ecotoxicological effects involving a series of bioassays on used snow will give more knowledge of the potential toxicity of different types of snow on the aquatic food chain.

References

ABQ 1990. *Symposium on the Saint-Laurence: a river to be reclaimed, comptes-rendus* (Proceedings). Edited by Messier, Legendre and Delisle. Vol. 11, Collection Environnement, Université de Montréal, 745 pages.

André, P., and C.E. Delisle. 1989. Modification des pratiques d'élimination des neiges usées pour diminuer l'apport de contaminants au Fleuve St-Laurent. Compte-rendu du Symposium sur le Saint-Laurent, 3, 4 et 5 nov. 1989, Montreal, Quebec, Canada. *Association des Biologistes du Québec, in Collection Environnement* Vol. 11, 745 pages, pp. 345-357, 1990.

APHA (American Public Health Association). 1989. *Standard methods for the examination of water and wastewater.* 1134 pages.

Delisle, C.E. 1993. Environmental impact of snow disposal in urban areas. Talk presented at a provincial conference on snow management in St-Georges, Quebec, 22 pages in french + figures.

Delisle, C.E. 1995. La gestion des neiges usées. *Routes et Transport* 24(4):20-27.

Delisle, C.E. et al. 1993. *Caractérisation des neiges usées en fonction de la densité résidentielle.* Rapport final à Ville LaSalle, Verdun et Lachine, CDT, EPM, 124 pages.

Delisle, C.E., M.-F. Lapointe, et A. Leduc. 1990. L'échantillonnage des neiges usées en milieu urbain: Résultats de l'hiver 1989-1990. *Sciences et Techniques de l'eau* 23(4):391-395.

Landsberger, S. 1984. Sulphur and heavy metal pollution of urban snow and environmental impact, Montreal, A.P.W.A. Snow Disposal Workshop, 8 p.

Landsberger, S. and P.E. Servis. 1983. Characterization of trace elemental pollutants in urban snow. *Int. J. of Env. Anal. Chem.* 16:95-130.

Leduc, A. and C.E. Delisle. 1990. Quality of used snow from city streets and sidewalks in Montreal, Quebec. Int'l Conf. on Urban Hydrology Under Wintry Conditions. Narvik, Norway (Winter Cities), 10 pages.

Lewis, J.E., T.R. Moore, and N.J. Enright. 1983. Spatial-temporal variations in smowfall chemistry in the Montreal region. *Water, Air and Soil Pollution* 20:7-22.

Loranger, S. and J. Zayed. 1995. Environmental contamination and human exposure to manganese: environmental fate/exposure modeling approach. In *1er Colloque de l'Université de Montréal sur l'environnement.* Edited by C. Delisle et al. Collection Environnement de l'Université de Montréal. Vol. 18:528-533.

MENVIQ. 1988. Politique sur l'élimination des neiges usées. Ministère de l'Environnement du Québec, 15 p. *Envirodoq* No. 880460.

Percherkin, N.A. and E.A. Burmatova. 1965. *Chemical composition of snow on the water surface of the Kama and Votkinst Reservoirs.* U.S. Army Cold Regions Research and Engineering Laboratory.

Pinard, D. et al. 1989. Caractérisation des eaux de fonte d'un dépôt à neiges usées. *Sc. et Techn. de l'eau* 22(3):211-215.

Riverin, J.M. 1982. L'élimination de la neige et la protection de l'environnement, Chicoutimi Env. Comity, 10 pages.

Scott, W. 1980. Occurence of salt and lead in snow dump sites. *Water, Air and Soil Pollution* 13:187-195.

Zinger, I. and C.E. Delisle. 1988. Quality of used-snow discharged in the St-Laurent River, in the region of the Montréal harbor. *Water, Air and Soil Pollution* 39:47-57.

Impact of Climate Change on Ecological Reclamation: A Study for Mackenzie Basin

by G.H. Huang and S.C. Cohen

Faculty of Engineering, University of Regina, Regina, Saskatchewan, and Sustainable Development Research Institute, University of British Columbia, Vancouver

Abstract. In this study, integrated climate-change impact assessment and adaptation analysis for ecological reclamation activities in the Mackenzie River Basin, Canada, were conducted through development/application of an interval mathematical programming (IMP) approach that can reflect uncertain, dynamic and interactive features of the study system. The Mackenzie Basin is situated in northwestern Canada and is the largest river basin in the country. The basin contains many different ecosystems (forest, wetland, delta, agriculture, etc.) and several important climate-sensitive boundaries including treelines and the southern extent of permafrost. It has essentially a natural resources based economy, and a number of ecological sectors are vulnerable to climate change. Thus, understanding the impacts of climate change on ecological reclamation is crucial to sustainable development of the region. The major activities under consideration are within agricultural/forest sectors since they are significantly related to climate change. Implications of climate change on achieving regional economic and resources development objectives are evaluated. Scenarios of ecological condition variation due to climate change are estimated and provided as intervals to represent a variety of alternatives. Through the proposed IMP approach, interactive relationships between different ecological reclamation activities and between different system objectives/constraints are effectively reflected. Potential conflicts and compromises between different system components are highlighted. Thus, alternatives corresponding to different strategies/policies for long-term adaptation planning of ecological reclamation activities in anticipation of global warming are generated. The results indicate that complex natures of climate-change impacts have been effectively reflected through the proposed approaches.

Introduction

For studies of regional climate-change impact assessment and adaptation strategy analysis, a number of biophysical and socioeconomic factors may have to be considered due to their direct or indirect relations to climate change (Yin and Cohen 1994). Integrated assessment that incorporates individual system components within a general framework rather than examining them in isolation may be useful for providing holistic and comprehensive analysis of climate-change impacts, as well as relevant policy responses for ecological reclamation of the existing life-support system (Dzidonu and Foster 1993; Yin et al. 1994).

In fact, many of the above system components may have uncertain features in practical problems. They may not be known with certainty but as follows: "the land area suitable for agricultural production is between 40 x 10^3 and 50 x 10^3 ha," "the benefit of forest timbering in the region is about $10/ha," and so on. Also, interactive relationships exist between different components, and may vary temporally with dynamic features (Dowlatabadi and Morgan 1993; Rotmans et al. 1994). For example, variations of climatic/economic/social conditions over time may lead to changes in, and thus conflicts between, agricultural/timbering activities, and may need compromises between different stakeholders in order to obtain an overall optimal situation; prices of agricultural products may affect the planning of crop production levels; expansion of crop production may have impacts on forest cover (and thus timber production) due to land use conflicts. These facts emphasize the need for a systematic approach for integrated assessment to effectively reflect the above uncertain, interactive, and dynamic features.

The objective of this paper is to report on research investigating integrated climate-change impact assessment and adaptive planning in the Mackenzie Basin, Canada, and developing an optimization/assessment approach for the study problem. An interval mathematical programming (IMP) model that can reflect the complex system features will be provided. As an example, a prototype case study of climate-change impact on agricultural/timber harvesting activities in the Mackenzie Basin will be described to demonstrate the applicability of the proposed methodologies. Implications of climate change on achieving regional economic and resources development objectives will be evaluated. The inclusion of agricultural/timber harvesting activities in this case is based mainly on the consideration of data availability at the current stage. In fact, many other sectors, such as wildlife habitat, wetland, and recreational activities, also can be affected by climate change and have interactions with the agricultural/timber harvesting activities. Scenarios of agricultural/forest land availability variation due to climate change will be estimated and provided as intervals to represent a variety of alternatives. The use of interval scenarios is based on the fact that many impact factors are uncertain, and that few studies have been reported regarding detailed impact scenarios for individual sectors in the basin area. Through the proposed IMP approach, interactive relationships between different land-use activities and between different system objectives/constraints will be effectively reflected. Potential conflicts and compromises between different system components will be highlighted. Thus, alternatives corresponding to different adaptive strategies and related risk levels can be generated through interpretation of the IMP outputs.

This study is part of the integrated phase of the Mackenzie Basin Impact Study (MBIS) which is a multiyear, multidisciplinary research project designed to examine regional impacts of projected global warming (Cohen 1994).

Overview of the Study System

The study area. The study area, the Mackenzie Basin, is situated in northwestern Canada and is the largest river basin in the country. The basin includes northeast British

Columbia, northern Alberta, northwest Saskatchewan, the Mackenzie Valley within the Northwest Territories, and portions of southwest and northern Yukon Territory, with an area of 1.79 million km^2 (Figure 1). It contains many different ecosystems (forest, wetland, delta, etc.) and several important climate-sensitive boundaries, including treelines and the southern extent of permafrost. The basin essentially has a natural resources-based economy, and a number of resource sectors are vulnerable to climate change (Cohen 1994). Understanding the impacts of climate change on the resource use system is crucial to sustainable development of the region. Due to complex and uncertain natures of the interactive relationships between different system components under changing climate, using an effective impact assessment approach within the MBIS seems desirable.

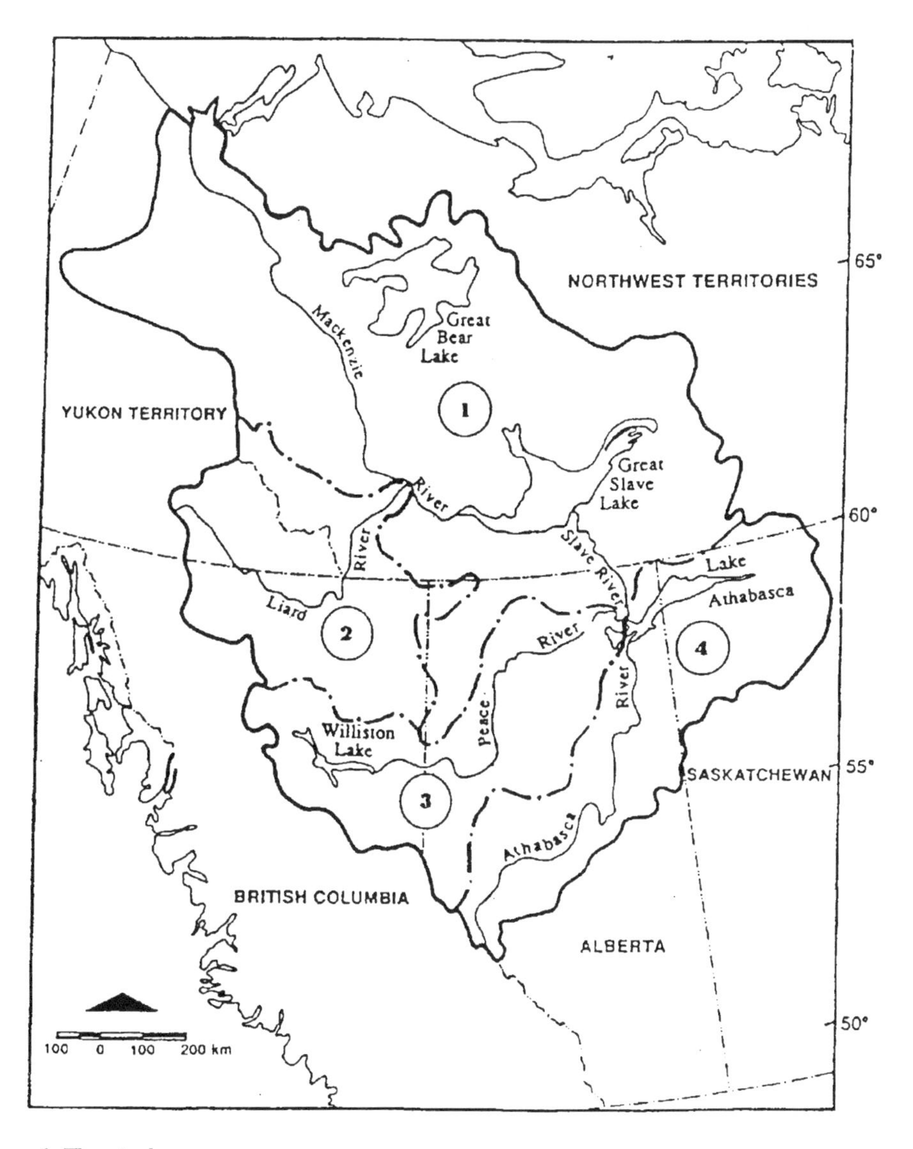

Figure 1. The study area.

Assumptions. In this paper, a problem is studied based on data availability. The major purpose of this case study is to describe and explain the IMP approach and its applicability to integrated climate change impact analysis. Generally, this research is based on the following assumptions:

(1) Only two sectors, including agriculture and timber harvesting, are considered in this study. Although the two sectors contain the major activities in the basin area, other sectors, such as wildlife habitat and recreational activities, also are important in the basin system. Therefore, this study should not be considered as a complete integration.

(2) Interval scenarios of climate change impacts on land availability for different activities were generated by an interval analysis approach based on the available uncertain information. It is expected that, after numerical outputs of scenarios (less uncertain) from individual MBIS sub-projects are generated, they then can be incorporated within this modelling framework to improve certainty of the IMP inputs and outputs. Figure 2 shows an example of potential relationships between interval scenarios and sub-project output scenarios.

(3) To reflect interregional and spatial considerations in the assessment, the basin was further divided into four subareas with different environmental, economic, and resources characteristics, corresponding to critical zones determined by the MBIS Working Committee (Figure 1) (Cohen 1994).

(4) The study time horizon was 30 years (1994 to 2024), which was further divided into three planning periods (each with an interval of 10 years). Over the 30-year planning horizon, it was assumed that global warming would lead to impacts on socioeconomic and biophysical sectors in the basin and affect agricultural and timber harvesting activities. A longer time horizon with shorter time interval can be considered in more detailed studies.

(5) The information was sorted into land-based activities. The main land use activities in the agricultural sector are assumed to include grain and hay production,

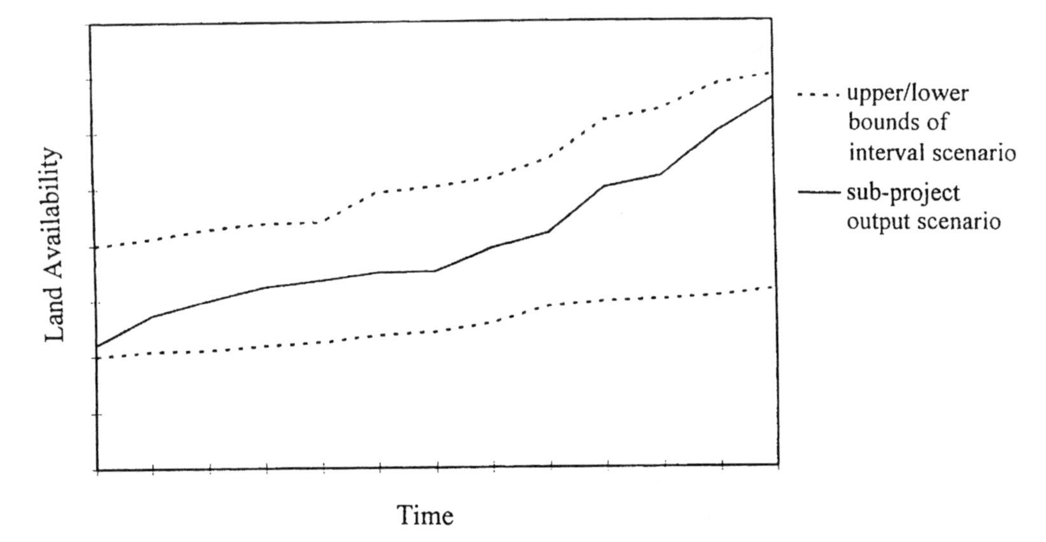

Figure 2. Relationships between interval scenarios and sub-project output scenarios.

and those in the forest sector to include spruce, pine, and aspen timber harvesting. These activities may correspond to different land units based on climatic and biophysical suitability.

Table 1 shows land area of different subareas in the basin. Table 2 presents land areas for different land use activities. More detailed discussion on the database can be found in Cohen (1993, 1994).

Statement of problems. For each time period with given climate-change scenario and environmental/economic conditions, there may be interactions and conflicts between different land use activities and between different system constraints/objectives. For example, climate change may lead to an increase in tillable land area for agriculture, and thus affect the planning of agricultural activities. Variation of prices for agricultural/forestry products may lead to changes in the planning of land use patterns. Also, agricultural and timber land uses may cause long-term environmental and ecological degradation problems and constraints/objectives. For each time period with given climate-change scenario and environmental/economic conditions, there may be interactions and conflicts between different land use activities and between different systems constraints/objectives. For example, climate change may lead to an increase in tillable land area for agriculture, and thus affect the planning of agricultural activities. Variation of prices for agricultural/forestry products may lead to changes in planning

Table 1. Land area of different subareas and subbasins

Subarea	Province/Territory	Subbasin	Land Area (1000 km^2)
1	Northwest Territories/Yukon Territory	Mackenzie River	346.0
1	Northwest Territories	Great Bear Lake	157.7
1	Northwest Territories/Alberta	Great Slave Lake	340.1
2	British Columbia/Northwest Territories/Yukon Territory	Liard River	281.5
3	British Columbia/Alberta	Peace River	303.6
4	Alberta/Saskatchewan	Lake Athabasca	297.1

Table 2. Land area for different land use activities (hectare)

Subarea	Subbasin	Grain	Hay	Spruce	Pine	Aspen
1	Mackenzie River	0	0	396268	43010	97552
1	Great Bear Lake	0	0	279853	21650	54095
1	Great Slave Lake	0	0	274570	33804	100954
2	Liard River	56260	26618	252652	110380	74048
3	Peace River	754712	447540	91796	65534	44692
4	Lake Athabasca	345596	346186	144753	83805	72218

and land use patterns. Also, agriculture/timber land uses may cause long-term environmental and ecological degradation problems and affect recreational, transportational, and wildlife activities. Therefore, the problems under consideration include the assessment of implications of climate change for achieving regional resource development objectives, and the selection of preferred strategies for the long-term adaptive planning of land use activities in anticipation of global warming. The analysis must be able to reflect the dynamic and interactive natures of the study system and the uncertain features in the majority of available data. The IMP provides such capabilities and will be applied for solving the above problems.

Modelling Approaches

Generally, changes in climatic conditions may lead to a series of impacts on environment, resource, and economic activities/objectives in different temporal/spatial units. There are also interactions between these activities/objectives. Therefore, development and application of suitable systems analysis approaches to integrate a variety of system components (objectives, constraints, and activities) within a general modelling framework will be useful for comprehensive impact assessments and generating desired adaptation strategies.

Climate change will affect the distribution of land suitability for different activities (e.g. potential and productivity for agricultural and timber industries) in the basin area. Thus, the desired distribution of land use activities should be adjusted to adapt the climate-change impacts and obtain highest possible environmental/economic benefits. The IMP is capable of yielding adequate information for the potential responses of agricultural/timber activities to climatic change. The IMP solutions can be compared to each other and to the present pattern in order to assess the impacts and finally generate desired adaptation strategies.

The dynamic impact assessment and adaptation planning problem under investigation may be broken up into individual stages for solution by integer programming, with these stages corresponding to the time intervals within the planning horizon. The entire basin, the entire time horizon (30 years), and the interactions between different components are integrated within a general system. The decision variables in the system include two categories: continuous and binary. The continuous variables represent agricultural/timber activities in different spatial locations over the time horizon, and the binary variables represent the dynamic characteristics of activity development decisions corresponding to the variations of land suitability due to climate change. Decision variables used in this study are not exclusive, since the model is flexible in incorporating other variables, such as wetland management intensity and wildlife habitat preservation, for assessment. The objective is to achieve optimal adaptation planning of activity development and relevant land use with maximized system benefit. The constraints include all relationships between decision variables and climate-change situations.

We first introduce some useful definitions. Let x denote a closed and bounded set of real numbers. An interval number $x^{\pm}$ is defined as an interval with known upper and lower bounds but unknown distribution information for x:

$$x^{\pm} = [x^{-}, x^{+}] = \{t \in x \mid x^{-} \le t \le x^{+}\}, \tag{1}$$

where x^{-} and x^{+} are the lower and upper bounds of $x^{\pm}$, respectively. When $x^{-} = x^{+}$, $x^{\pm}$ becomes a deterministic number.

Letting $* \in \{+, -, \times, \div\}$ be a binary operation on interval numbers, we have the following for $x^{\pm}$ and $y^{\pm}$:

$$x^{\pm} * y^{\pm} = [\min\{x * y\}, \max\{x * y\}], \quad x^{-} \le x \le x^{+}, y^{-} \le y \le y^{+}. \tag{2}$$

In the case of division, it is assumed that $y^{\pm}$ does not contain a zero.

Also, for $x^{\pm}$ and $y^{\pm}$, we have their order relations as follows:

$$x^{\pm} \le y^{\pm}, \text{iff } x^{-} \le y^{-} \text{and } x^{+} \le y^{+},. \tag{3}$$

$$x^{\pm} < y^{\pm}, \text{iff } x^{-} < y^{-} \text{and } x^{+} < y^{+}. \tag{4}$$

Thus an interval mixed integer linear programming (IMILP) model for the dynamic optimization problem can be formulated as follows:

$$\max\ f^{\pm} = \sum_{i=1}^{5} \sum_{j=1}^{4} \sum_{k=1}^{3} [(B^{\pm}_{ijk} - D^{\pm}_{ijk})x^{\pm}_{ijk} - \sum_{m=1}^{2} C^{\pm}_{ijmk}\, y^{\pm}_{ijmk}], \tag{5a}$$

[objective function]

$$\text{s.t.} \quad \sum_{i=1}^{2} x^{\pm}_{ijk'} \le \sum_{i=1}^{2} \sum_{m=1}^{2} \sum_{k=1}^{k'} \Delta x^{\pm}_{ijm}\, y^{\pm}_{ijmk} + P^{\pm}_{1j},\ k' = 1,\ 2,\ 3,\ \forall j, \tag{5b}$$

[constraints of land availability for agricultural production]

$$\sum_{i=3}^{5} x^{\pm}_{ijk'} \le \sum_{i=3}^{5} \sum_{m=1}^{2} \sum_{k=1}^{k'} \Delta x^{\pm}_{ijm}\, y^{\pm}_{ijmk} + P^{\pm}_{2j},\ k' = 1,\ 2,\ 3,\ \forall j, \tag{5c}$$

[constraints of land availability for timbering production]

$$x^{\pm}_{ijk} \le R^{\pm}_{ijk},\ \forall\ i,\ j,\ k, \tag{5d}$$

$$x^{\pm}_{ijk} \ge Q^{\pm}_{ijk},\ \forall\ i,\ j,\ k, \tag{5e}$$

[constraints of agricultural/timbering production levels]

$$\sum_{m=1}^{2} y^{\pm}_{ijmk} \le\ 1,\ \forall\ i,\ j,\ k,\ (5f)$$

[only one activity expansion may occur for any given subarea/period]

$$y^{\pm}_{ijmk} \le 1,$$
$$\ge 0,$$
$$= \text{integer},\ \forall\ i,\ j,\ m,\ k, \tag{5g}$$

[non-negativity and binary constraints]

$$x^{\pm}_{ijk} \ge 0,\ \forall\ i,\ j,\ k, \tag{5h}$$

[non-negativity constraints]

where:

$B^{\pm}_{ijk}$ = benefit from activity i in subarea j during period k;

$C^{\pm}_{ijmk}$ = cost for expansion of activity i with increment m in subarea j during period k;

$D^{\pm}_{ijk}$ = cost/loss from activity i in subarea j during period k;

$f^{\pm}$ = net system benefit;

i = land use activity (i = 1 for grain, 2 for hay, 3 for spruce, 4 for pine, and 5 for aspen);

j = subarea, where j = 1 for Mackenzie River, Great Bear Lake and Great Slave Lake Basins, 2 for Liard River Basin, 3 for Peace River Basin, and 4 for Lake Athabasca Basin;

k = time period;

m = expansion option;

$P^{\pm}_{1j}$ = land availability for agricultural production in subarea j;

$P^{\pm}_{2j}$ = land availability for timbering production in subarea j;

$Q^{\pm}_{ijk}$ = least allowable land use for activity i in subarea j during period k;

$R^{\pm}_{ijk}$ = highest allowable land use for activity i in subarea j during period k;

$x^{\pm}_{ijk}$ = land use for activity i in subarea j during period k;

$y^{\pm}_{ijmk}$ = 0-1 decision variable for expansion of activity i with increment m in subarea j during period k;

$\Delta x^{\pm}_{im}$ = amount of expansion for activity i with increment m.

Equation (5a) means that the objective is to maximize net system benefit which is based on the consideration of benefits/costs from different land use activities and capital costs for related activity expansions/developments. Constraints (5b) and (5c) mean that land availability for agricultural/timber production in any time stage is related to both existing and expanded/developed lands. This dynamic nature can be caused by climate change and/or economic consideration. Constraints (5d) and (5e) are limitations of agricultural/timber production levels in different spatial/temporal units. They can be defined based on the consideration of ecological sustainability and environmental-economic objectives. Constraint (5f) means that only one activity expansion may occur for any given subarea/period based on the principle of economies of scale, while (5g) and (5h) are technical constraints for the decision variables.

A solution algorithm for the above IMILP problem was provided by Huang et al. (1995), with interval solutions as follows:

$$x^{\pm}_{ijk\,opt} = [x^{-}_{ijk\,opt}\,,\ x^{+}_{ijk\,opt}],\ \forall\ i,\ j,\ k, \tag{6}$$

$$y^{\pm}_{ijmk\,opt} = [y^{-}_{ijmk\,opt}\,,\ y^{+}_{ijmk\,opt}],\ \forall\ i,\ j,\ m,\ k, \tag{7}$$

$$f^{\pm}_{opt} = [f^{-}_{opt}\,,\ f^{+}_{opt}]. \tag{8}$$

Among the variables/parameters in model (5), $R^{\pm}_{ijk}$, $Q^{\pm}_{ijk}$, $x^{\pm}_{ijk}$ and $y^{\pm}_{ijmk}$ are sensitive to climate change. For example, climate change may lead to increased land suitability for agriculture, which may potentially lead to increase in relevant $R^{\pm}_{ijk}$ and $y^{\pm}_{ijmk}$ values.

The coefficients in (5a) are expressed in present value dollars. They are escalated to reflect anticipated conditions and then discounted to generate present value terms for the objective function. For example, assume that the future values for $C^{\pm}{}_{ijmk}$ are $F^{\pm}{}_{ijmk}$. We have the relation between $C^{\pm}{}_{ijmk}$ and $F^{\pm}{}_{ijmk}$ as follows:

$$C^{\pm}{}_{ijmk} = F^{\pm}{}_{ijmk}\ (1+i)^{-10\,(k-1)}, \tag{9}$$

where i is a discount rate.

The IMILP model allows uncertainty to be directly communicated into the optimization process and provides solutions presented as intervals. The modelling results can be used to generate decision alternatives by adjusting different combinations of decision variable values within the solution intervals, and thus answer a number of questions in the initial study stages and throughout the study horizon. The feasible ranges for decision variables provided by the IMILP solutions also are useful for decision makers to justify the generated alternatives directly, or to potentially adjust adaptation schemes when they are not satisfied with the provided alternatives.

The IMILP also has reasonable computational requirements since its solution algorithm does not lead to more complicated intermediate models. Moreover, the method does not require distribution information since interval numbers are used to represent uncertain inputs/outputs. This is particularly meaningful for practical applications because (1) it is typically more difficult to specify distributions than to define fluctuation intervals, and (2) the existing optimization methods that deal with distribution uncertainties have difficulties in solution algorithms and computational requirements. Certainly, the IMILP model is also flexible in incorporating stochastic and fuzzy variables where the data support this level of precision (Huang et al. 1994 and 1995).

Results

General pattern. Table 3 shows solutions of the IMILP model in comparison with current land use patterns. They provide schemes for agricultural/timber activities in different subareas/periods. A number of model outputs are present as intervals, which reflect the effects of input uncertainties. Generally, it is indicated that the solutions for period 1 do not differ significantly from the current pattern. The only exception is for grain production, which actually is affected by many other factors in addition to climatic conditions. This result means that little impact can be found from the short-term point of view, and thus little effort is required to adapt the potential changes in period 1.

However, as time goes on, the differences between the periods become more and more significant. For example, from period 1 to 3, land for hay, aspen, and pine should be significantly increased, and that for spruce should be decreased. In comparison, the temporal variation of grain production level during the 30-year horizon is quite uncertain (Figure 3).

Figure 4 shows a more detailed description of temporal/spatial variations of each activity. It is indicated that the majority of hay, aspen, and pine related activities would be increased along with time, which fits the general trends shown in Figure 3. Land for spruce production generally should be decreased. For grain production, the cropping area for subarea 2 should be slightly decreased, that for subarea 4 should be slightly increased, and that for subarea 3, which contributes the most to grain production in the basin, is uncertain with regard to trend. This uncertainty is due to the fact

Table 3. Solutions of IMILP model					
$x^{\pm}_{ijk\ opt}$	**Activity**	**Subarea**	**Period**	**Solution**	**Expansion (10^6 ha)**
Binary variables (only non-zero solutions are presented):					
$y^{\pm}_{1413opt}$	grain	4	3	1	[0.050, 0.063]
$y^{\pm}_{2313opt}$	hay	3	3	1	[0.43, 0.48]
$y^{\pm}_{4213opt}$	pine	2	3	1	[0.025, 0.034]
$y^{\pm}_{5113opt}$	aspen	1	3	1	[0.057, 0.070]
Current land use pattern (10^6 ha):					
	grain	1	0	0	
	grain	2	0	0.056	
	grain	3	0	0.75	
	grain	4	0	0.35	
	hay	1	0	0	
	hay	2	0	0.027	
	hay	3	0	0.45	
	hay	4	0	0.35	
	spruce	1	0	0.95	
	spruce	2	0	0.25	
	spruce	3	0	0.092	
	spruce	4	0	0.14	
	pine	1	0	0.098	
	pine	2	0	0.11	
	pine	3	0	0.066	
	pine	4	0	0.084	
	aspen	1	0	0.25	
	aspen	2	0	0.074	
	aspen	3	0	0.045	
	aspen	4	0	0.072	
Variables for period 1 (10^6 **ha**):					
$x^{\pm}_{111opt}$	grain	1	1	0	
$x^{\pm}_{121opt}$	grain	2	1	0.19	
$x^{\pm}_{131opt}$	grain	3	1	[1.55, 1.57]	
$x^{\pm}_{141opt}$	grain	4	1	[0.12, 0.15]	
$x^{\pm}_{211opt}$	hay	1	1	0	
$x^{\pm}_{221opt}$	hay	2	1	[0.028, 0.032]	
$x^{\pm}_{231opt}$	hay	3	1	0.53	
$x^{\pm}_{241opt}$	hay	4	1	[0.34, 0.44]	
$x^{\pm}_{311opt}$	spruce	1	1	[0.9, 1.35]	
$x^{\pm}_{321opt}$	spruce	2	1	0.24	
$x^{\pm}_{331opt}$	spruce	3	1	[0.103, 0.110]	
$x^{\pm}_{341opt}$	spruce	4	1	[0.19, 0.25]	

Table 3. Solutions of IMILP model (continued)					
$x^{\pm}_{ijk\ opt}$	**Activity**	**Subarea**	**Period**	**Solution**	**Expansion 10^6 ha)**
Variables for period 1 (10^6 **ha) (continued):**					
$x^{\pm}_{411opt}$	pine	1	1	[0.12, 0.16]	
$x^{\pm}_{421opt}$	pine	2	1	[0.14, 0.19]	
$x^{\pm}_{431opt}$	pine	3	1	[0.07, 0.09]	
$x^{\pm}_{441opt}$	pine	4	1	0.07	
$x^{\pm}_{511opt}$	aspen	1	1	[0.26, 0.27]	
$x^{\pm}_{521opt}$	aspen	2	1	[0.09, 0.15]	
$x^{\pm}_{531opt}$	aspen	3	1	[0.018, 0.050]	
$x^{\pm}_{541opt}$	aspen	4	1	[0.059, 0.068]	
Variables for period 2 (10^6 ha):					
$x^{\pm}_{112opt}$	grain	1	2	0	
$x^{\pm}_{122opt}$	grain	2	2	0.14	
$x^{\pm}_{132opt}$	grain	3	2	[1.38, 2.07]	
$x^{\pm}_{142opt}$	grain	4	2	0.15	
$x^{\pm}_{212opt}$	hay	1	2	[0.06, 0.11]	
$x^{\pm}_{222opt}$	hay	2	2	[0.033, 0.040]	
$x^{\pm}_{232opt}$	hay	3	2	0.70	
$x^{\pm}_{242opt}$	hay	4	2	[0.40, 0.50]	
$x^{\pm}_{312opt}$	spruce	1	2	[0.8, 1.25]	
$x^{\pm}_{322opt}$	spruce	2	2	[0.22, 0.28]	
$x^{\pm}_{332opt}$	spruce	3	2	0.74	
$x^{\pm}_{342opt}$	spruce	4	2	[0.16, 0.20]	
$x^{\pm}_{412opt}$	pine	1	2	[0.18, 0.25]	
$x^{\pm}_{422opt}$	pine	2	2	0.17	
$x^{\pm}_{432opt}$	pine	3	2	[0.12, 0.13]	
$x^{\pm}_{442opt}$	pine	4	2	[0.08, 0.10]	
$x^{\pm}_{512opt}$	aspen	1	2	[0.29, 0.39]	
$x^{\pm}_{522opt}$	aspen	2	2	0.12	
$x^{\pm}_{532opt}$	aspen	3	2	[0.042, 0.079]	
$x^{\pm}_{542opt}$	aspen	4	2	[0.062, 0.066]	
Variables for period 3 (10^6 ha):					
$x^{\pm}_{113opt}$	grain	1	3	0	
$x^{\pm}_{123opt}$	grain	2	3	[0.09, 0.12]	
$x^{\pm}_{133opt}$	grain	3	3	[1.56, 2.00]	
$x^{\pm}_{143opt}$	grain	4	3	[0.18, 0.22]	
$x^{\pm}_{213opt}$	hay	1	3	0.10	
$x^{\pm}_{223opt}$	hay	2	3	[0.040, 0.049]	
$x^{\pm}_{233opt}$	hay	3	3	[0.90, 1.40]	
$x^{\pm}_{243opt}$	hay	4	3	[0.48, 0.56]	

Table 3. Solutions of IMILP model (continued)					
$x^{\pm}_{ijk\ opt}$	**Activity**	**Subarea**	**Period**	**Solution**	**Expansion 10^6 ha)**
Variables for period 3 (10^6 ha) (continued):					
$x^{\pm}_{313opt}$	spruce	1	3	[0.74, 0.77]	
$x^{\pm}_{323opt}$	spruce	2	3	0.16	
$x^{\pm}_{333opt}$	spruce	3	3	[0.039, 0.045]	
$x^{\pm}_{343opt}$	spruce	4	3	0.11	
$x^{\pm}_{413opt}$	pine	1	3	[0.26, 0.30]	
$x^{\pm}_{423opt}$	pine	2	3	[0.20, 0.26]	
$x^{\pm}_{433opt}$	pine	3	3	0.18	
$x^{\pm}_{443opt}$	pine	4	3	[0.10, 0.12]	
$x^{\pm}_{513opt}$	aspen	1	3	[0.33, 0.42]	
$x^{\pm}_{523opt}$	aspen	2	3	0.15	
$x^{\pm}_{533opt}$	aspen	3	3	0.070	
$x^{\pm}_{543opt}$	aspen	4	3	[0.067, 0.076]	
$f^{\pm}_{opt}$ (\$$10^6$)				[62.3, 120.1]	

that grain production is affected by many environmental/economic factors in addition to climate change. For spatial variation, activities in subareas 1 and 3 have more significant temporal variations than those in subareas 2 and 4.

The results demonstrate that agricultural/timber activities should be adjusted to adapt to temporal variations of climatic/economic conditions that are responsible for changes in the structure and dynamics of land availability. These activity variations also are different from each other for the four subareas due to differences in local economic, resource, and environmental conditions. Figure 5 presents temporal variation

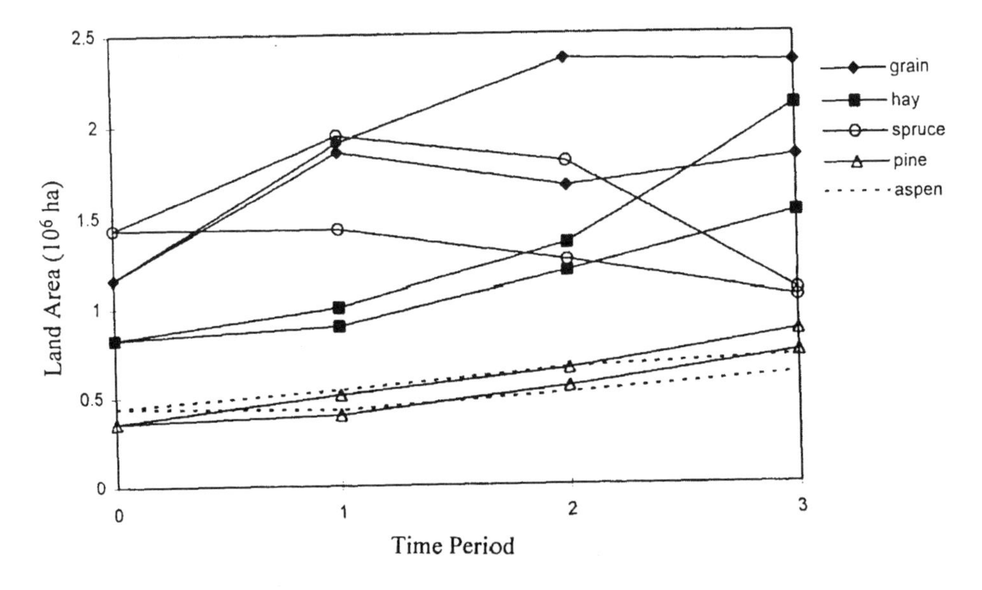

Figure 3. Temporal variations of each land use activity.

of the total of all five activities in each subarea. It is indicated that a significant increase of land use with time can be found in subarea 3, while variations in subareas 1, 2 and 4 are not very significant. The major reason for the lack of significance in subareas 1, 2, and 4 is that, although individual activities can have significant temporal variations, they may interact with each other and finally offset the general variation in a subarea.

Alternatives for adaptation strategy. When the detailed adaptation strategies are to be determined, decision makers prefer that a set of alternatives can be provided. The interval outputs generated by the IMILP model can be interpreted to provide such alternatives. Table 4 presents four alternatives obtained by adjusting $x^{\pm}_{ijk}$ values within their solution intervals. It is indicated that both agricultural and timber activities reach

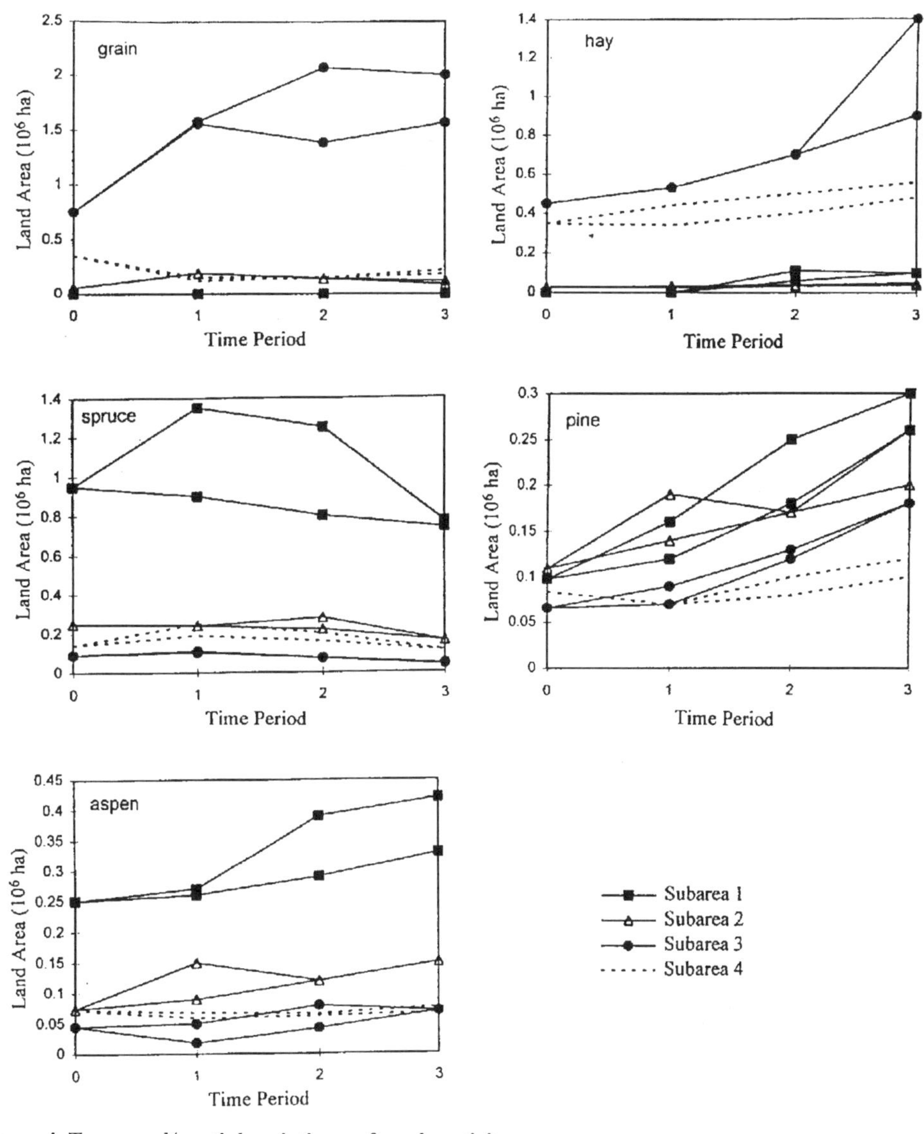

Figure 4. Temporal/spatial variations of each activity.

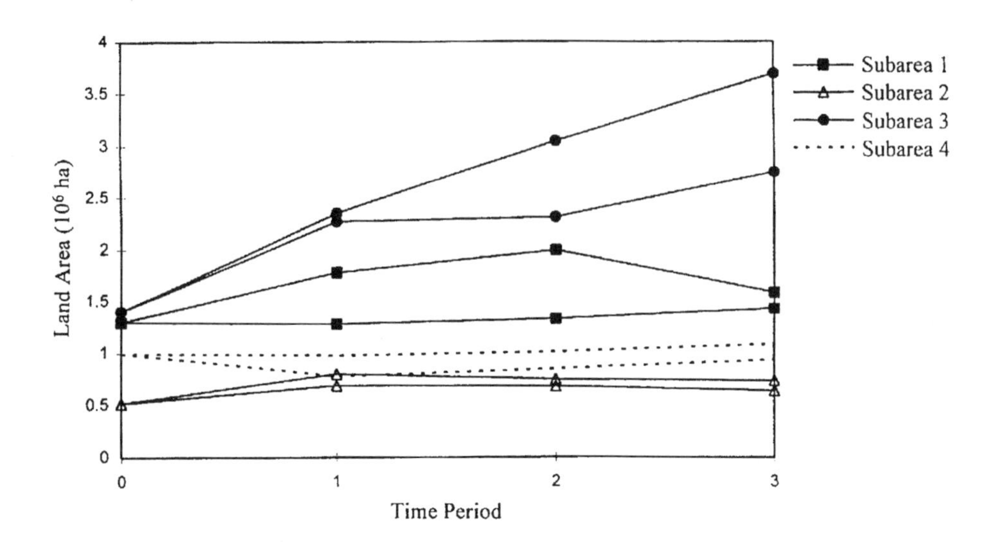

Figure 5. Temporal variations in each subarea.

their upper bounds in Alternative 1. Under this alternative, higher system benefit may be obtained with higher risk of conflict with other land use sectors (and thus higher risk of poor adaptation). In Alternative 2, all agricultural/timber activities fall towards the lower bounds, which may lead to lower benefits with lower risk of poor adaptation. Alternative 3 is suitable to the situation when more agricultural land use is allowed under changed climatic, environmental, and economic conditions, while Alternative 4 corresponds to more timber land use with less agricultural potential.

The alternatives in Table 4 were obtained by examining some combinations of the upper/lower bounds of the decision variables within their solution intervals. In practical implementation the variable values can be adjusted continuously within their solution intervals, which allows managers to more conveniently incorporate implicit within the problem, and thus obtain more satisfactory and applicable strategies.

Characteristics of the IMILP approach. (1) The IMILP modelling results provide bases for comparisons of optimized land use patterns in different periods. As shown in Figure 6, land availability for different activities will change with time due to variation of climatic and economic conditions. These temporal changes can be assessed directly. However, for impacts on land use activities, it is difficult to make direct assessments based on the existing patterns since pattern activities remain uncertain. The IMILP approach allows this type of assessment based on the generated solutions for agricultural/timber activities in periods 1 to 3. These solutions provide most desirable patterns for different time stages, and thus form a common ground for effective comparison.

(2) Uncertainties in climate change scenarios were quantified through interval numbers. This reflects the fact of data availability and quality, and the related information can be incorporated within the IMILP model.

(3) Through the proposed modelling approach, decision alternatives can be generated and interpreted for internalizing tradeoffs between different system objectives. This

Table 4. Alternatives obtained from IMILP solutions

$x^{\pm}_{ijk\ opt}$	Activity	Alternative 1	Alternative 2	Alternative 3	Alternative 4
Variables for period 1					
$x^{\pm}_{111opt}$	grain	+	–	+	–
$x^{\pm}_{121opt}$	grain	+	–	+	–
$x^{\pm}_{131opt}$	grain	+	–	+	–
$x^{\pm}_{141opt}$	grain	+	–	+	–
$x^{\pm}_{211opt}$	hay	+	–	+	–
$x^{\pm}_{221opt}$	hay	+	–	+	–
$x^{\pm}_{231opt}$	hay	+	–	+	–
$x^{\pm}_{241opt}$	hay	+	–	+	–
$x^{\pm}_{311opt}$	spruce	+	–	–	+
$x^{\pm}_{321opt}$	spruce	+	–	–	+
$x^{\pm}_{331opt}$	spruce	+	–	–	+
$+^{+}_{341opt}$	spruce	+	–	–	+
$x^{\pm}_{411opt}$	pine	+	–	–	+
$x^{\pm}_{421opt}$	pine	+	–	–	+
$x^{\pm}_{431opt}$	pine	+	–	–	+
$x^{\pm}_{441opt}$	pine	+	–	–	+
$x^{\pm}_{511opt}$	aspen	+	–	–	+
$x^{\pm}_{521opt}$	aspen	+	–	–	+
$x^{\pm}_{531opt}$	aspen	+	–	–	+
$x^{\pm}_{541opt}$	aspen	+	–	–	+
Variables for period 2					
$x^{\pm}_{112opt}$	grain	+	–	+	–
$x^{\pm}_{122opt}$	grain	+	–	+	–
$x^{\pm}_{132opt}$	grain	+	–	+	–
$x^{\pm}_{142opt}$	grain	+	–	+	–
$x^{\pm}_{212opt}$	hay	+	–	+	–
$x^{\pm}_{222opt}$	hay	+	–	+	–
$x^{\pm}_{232opt}$	hay	+	–	+	–
$x^{\pm}_{242opt}$	hay	+	–	+	–
$x^{\pm}_{312opt}$	spruce	+	–	–	+
$x^{\pm}_{322opt}$	spruce	+	–	–	+
$x^{\pm}_{332opt}$	spruce	+	–	–	+
$x^{\pm}_{342opt}$	spruce	+	–	–	+
$x^{\pm}_{412opt}$	pine	+	–	–	+
$x^{\pm}_{422opt}$	pine	+	–	–	+
$x^{\pm}_{432opt}$	pine	+	–	–	+
$x^{\pm}_{442opt}$	pine	+	–	–	+
$x^{\pm}_{512opt}$	aspen	+	–	–	+
$x^{\pm}_{522opt}$	aspen	+	–	–	+

Table 4. Alternatives obtained from IMILP solutions (continued)

$x^{\pm}_{ijk\ opt}$	Activity	Alternative 1	Alternative 2	Alternative 3	Alternative 4
Variables for period 2 (continued):					
$x^{\pm}_{532opt}$	aspen	+	–	–	+
$x^{\pm}_{542opt}$	aspen	+	-	-	+
Variables for period 3:					
$x^{\pm}_{113opt}$	grain	+	–	+	–
$x^{\pm}_{123opt}$	grain	+	–	+	–
$x^{\pm}_{133opt}$	grain	+	–	+	–
$x^{\pm}_{143opt}$	grain	+	–	+	–
$x^{\pm}_{213opt}$	hay	+	–	+	–
$x^{\pm}_{223opt}$	hay	+	–	+	–
$x^{\pm}_{233opt}$	hay	+	–	+	–
$x^{\pm}_{243opt}$	hay	+	–	+	–
$x^{\pm}_{313opt}$	spruce	+	–	–	+
$x^{\pm}_{323opt}$	spruce	+	–	–	+
$x^{\pm}_{333opt}$	spruce	+	–	–	+
$+^{+}_{343opt}$	spruce	+	–	–	+
$x^{\pm}_{413opt}$	pine	+	–	–	+
$x^{\pm}_{423opt}$	pine	+	–	–	+
$x^{\pm}_{433opt}$	pine	+	–	–	+
$x^{\pm}_{443opt}$	pine	+	–	–	+
$x^{\pm}_{513opt}$	aspen	+	–	–	+
$x^{\pm}_{523opt}$	aspen	+	–	–	+
$x^{\pm}_{533opt}$	aspen	+	–	–	+
$x^{\pm}_{543opt}$	aspen	+	–	–	+
Objective function value:					
$f^{\pm}_{opt}$ ($\$10^6$)		[112.3, 120.1]	[60.3, 64.5]	[103.3, 108.9]	[73.1, 78.8]

Note: "+" = upper bound value, and "–" = lower bound value.

IMILP solution feature may be favoured by decision makers faced with difficult and controversial choices because of the increased flexibility and applicability for determining the final decision strategies.

Conclusions

An integrated climate-change impact assessment and adaptation study for sustainable ecological reclamation in the Mackenzie River Basin of Canada, was conducted recently through application of an IMP approach that can reflect complex system features. The results indicate that uncertain, dynamic, and interactive features of the study system have been effectively reflected. Generally, temporal variations of land

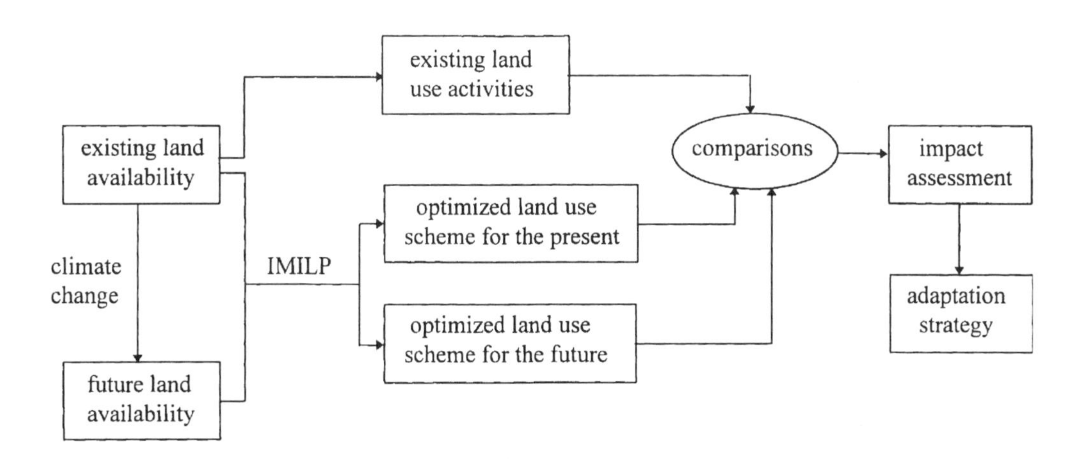

Figure 6. Land availability and land use

characteristics and thus land use activities exist due to changes in climatic, economic, and environmental conditions.

The IMILP can reflect not only particular structure and entities of a complex system, but also processes, interactions, and feedback mechanisms within the system that generates changes in its dynamics and structure. The method allows uncertainties to be effectively communicated into the optimization process and resulting solutions. It also has low computational requirements and is applicable to large-scale practical problems.

Acknowledgment

This research has been supported by Environment Canada and the Natural Science and Engineering Research Council of Canada.

References

Cohen, S.J., editor. 1993. *Mackenzie Basin impact study.* Interim Report #1, Canadian Climate Center, Downsview, Ontario, Canada.

Cohen, S.J., editor. 1994. *Mackenzie Basin impact study.* Interim Report #2, Environmental Adaptation Research Group, Environment Canada, Downsview, Ontario, Canada.

Dowlatabadi, H., and G.M. Morgan. 1993. Integrated assessment of climate change. *Science* 259:1813-1814.

Dzidonu, C.K., and F.G. Foster. 1993. Prolegomena to OR modelling of the global environment-development problem. *Journal of the Operational Research Society* 44: 321-33.

Huang, G.H., B.W. Baetz, and G.G. Patry. 1994. A chance-constrained programming approach for waste management planning under uncertainty. In *Effective environmental management for sustainable development.*. Edited by K.W. Hipel and L. Fang. Kluwer Academic Publishers, Dordrecht, The Netherlands, pp. 267-280.

Huang, G.H., B.W. Baetz, and G.G. Patry. 1995. Grey fuzzy integer programming: application to regional solid waste management planning. *Socio-Economic Planning Sciences* 29:17-38.

Rotmans, J., M.B.A. van Asselt, A.J. de Bruin, M.G.J. den Elzen, and J. de Greef. *Global change and sustainable development — a modelling perspective for the next decade*. GLOBO Report Series no. 4, National Institute of Public Health and Environmental Protection, Bilthoven, The Netherlands.

Yin, Y.Y., and S.J. Cohen. 1994. Identifying regional goals and policy concerns associated with global climate change. *Global Environmental Change* 4:246-260.

Yin, Y.Y., G.H. Huang, and S.J. Cohen. 1994. Designing an integrated climate change impact assessment framework for the Mackenzie River Basin in Canada. in *Proceedings of Decision Support - 2001*, Toronto, Ontario, Canada.

Atlantic Salmon (*Salmo salar* L.) Stock Recovery in the Gander River, Newfoundland, with Projections to 1999

by P. M. Ryan[1], R. Knoechel[2], M. F. O'Connell[1], E. G. M. Ash[1], and W. G. Warren[1]

1. Science Branch, Department of Fisheries and Oceans, P. O. Box 5667, St. John's, Nfld. A1C 5X1
2. Biology Department, Memorial University of Newfoundland, St. John's, Nfld. A1B 3X9

This paper is based upon a DFO stock assessment document of the same name: Ryan, P. M., R. Knoechel, M.F. O'Connell, E.G.M. Ash , and W.G. Warren. 1995. Atlantic salmon (*Salmo salar* L.) stock recovery in the Gander River, Newfoundland with projections to 1999. DFO Atlantic Fisheries Research Document 95/95. 16 p.

Abstract. Atlantic salmon returning as adults to spawn in the Gander River system of insular Newfoundland have been in low abundance, particularly during 1989-1991, resulting in an egg deposition of only about 35% of the target spawning requirement. Changes in the fisheries for adult salmon have included, starting in 1992, a closure of the commercial salmon fishery on the island and imposition of a quota in the recreational fishery (with subsequent catch and release fishing). In this paper we examine two adult-juvenile relationships, with data available up to 1994, for the purpose of assessing stock recovery in the Gander River system. We describe a numerical relationship between adult small salmon (<63 cm) returning to the Salmon Brook fishway on a lower tributary of the Gander River system and the juveniles in two lakes at the headwaters of the system (as recruits) four years later. For the first time we examine the numerical relationship between those juveniles and the adults returning to the entire river system one year later (as determined from a counting fence and angler survey on the main stem of the river since 1989). Subsequently, we estimate juvenile abundance from 1995 to 1998 from the stock-recruit relationship and then project total river adult small salmon returns from 1995 to 1999 from the post-commercial fishery ratio of total returning adults to the juvenile abundance one year earlier. Our results indicate that the number of salmon returning to the Gander River to spawn will not exceed the spawning requirement in 1995 or 1996 but should exceed the requirement from 1997-1999. Maximum returns over that period are calculated as 33,276 small salmon in 1998. Accordingly, we anticipate that the substantial reduction in fishing effort on this stock will result in recovery after 1996.

Introduction

The Gander River (Figure 1) has insular Newfoundland's largest river basin area that is naturally accessible to sea-run Atlantic salmon (Murray and Harmon 1969).

During the past 20 years, commercial marine catches of returning sea-run adults around the mouth of the river in Gander Bay have increased substantially, followed by a marked decline after 1989 (Figure 2). Concurrently, escapement of spawners to the river, as indicated by freshwater angling success rates, has plummeted (Figure 3). The target spawning requirement for the Gander River, or the number of spawning salmon required for sustained production, has been estimated as 21,828 small salmon (O'Connell and Dempson 1991). However, the decreased abundance of adults returning to spawn in the river has resulted in an egg deposition of only about 35% of that requirement, particularly during 1989-91 (Porter and O'Connell 1992; O'Connell and Ash 1993).

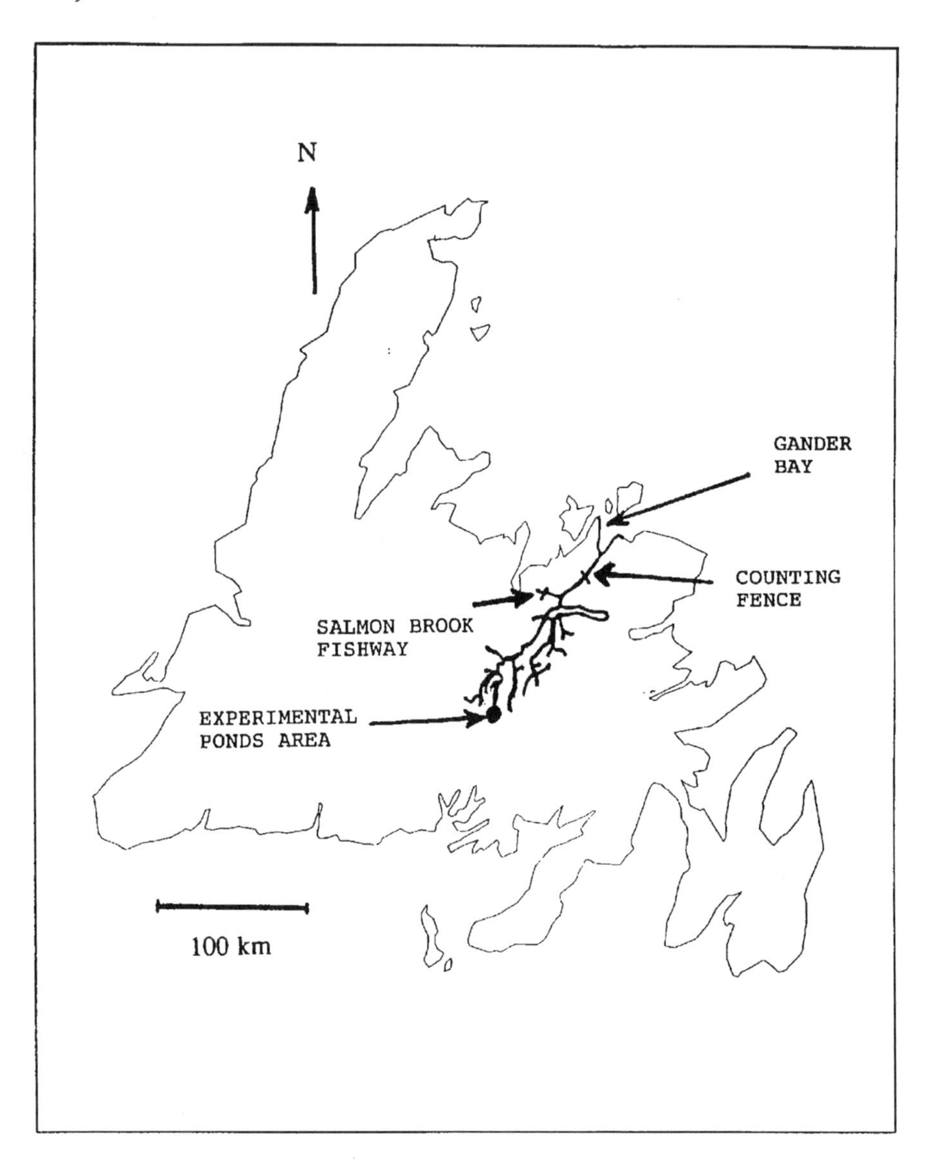

Figure 1. Gander River basin of insular Newfoundland with locations of study sites referred to in the text.

Changes in the fisheries for adult salmon have included, starting in 1992, a closure of the commercial salmon fishery on the island and the imposition of a quota in the recreational fishery (with subsequent catch and release fishing) (O'Connell and Ash 1993). As a result of the changing salmon fisheries and resultant changes in applicable fishing effort statistics, alternate methods to assess stock reclamation in the Gander River system have been explored. One of these has been the evaluation of variations

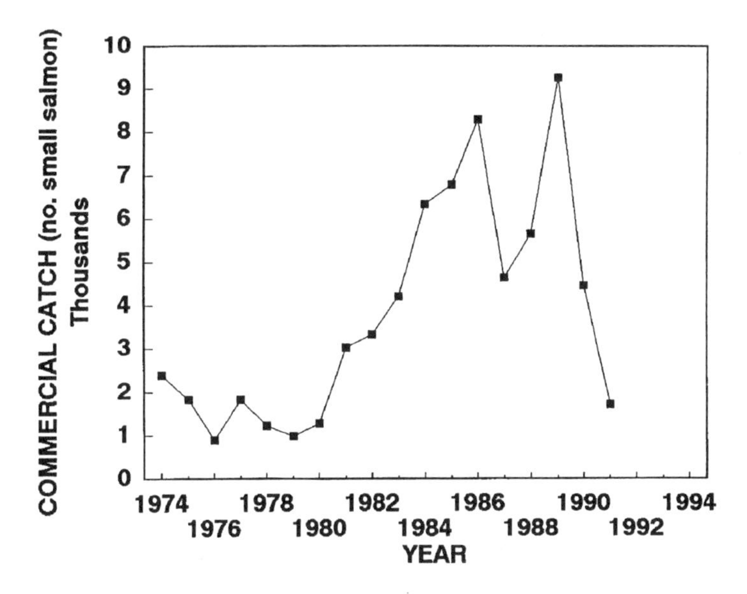

Figure 2. Commercial catch of small salmon (returning sea-run adults less than 63 cm in length) in Gander Bay at the mouth of the Gander River, insular Newfoundland. Historically, small salmon have represented in excess of 90% of all adults escaping to the river. For additional numerical detail, see Porter and O'Connell (1992).

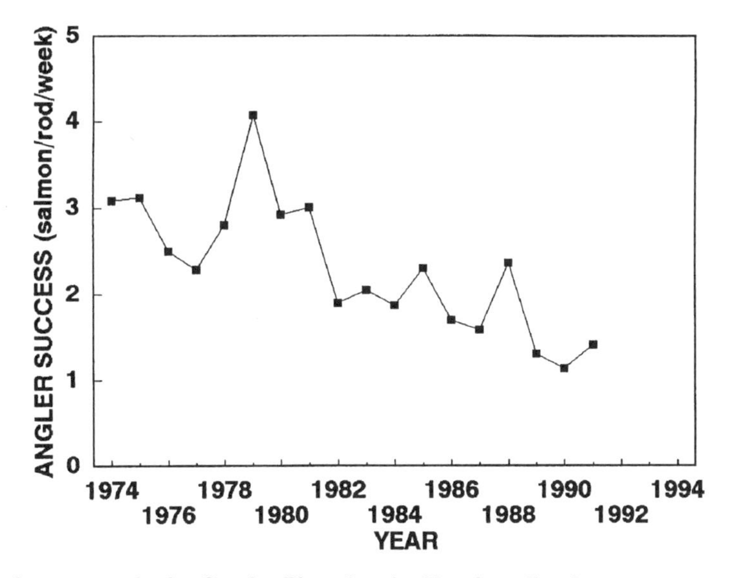

Figure 3. Angler success in the Gander River, insular Newfoundland.

in juvenile abundance with the assumption that increases in the abundance of juveniles are, unless shown otherwise, indicative of increases in the size of the spawning stock. Additionally, it has been expected that a stock-recruit relationship (described by Ryan, Knoechel, and O'Connell 1994) developed from adult escapement to the Salmon Brook fishway on a lower tributary of the system (as spawners) and the known abundance of juveniles in two lakes in the Experimental Ponds Area (EPA) at the headwaters of the Gander River (as recruits), would serve as a means of evaluating stock recovery. A positive relationship between adult escapement to the Salmon Brook fishway and angler success (fish/rod/week) throughout the entire river system (Figure 4) prior to the modification to the sport fishery has suggested that the Salmon Brook fishway count can provide an index of escapement to the entire river system.

In this paper we examine two adult-juvenile relationships with data available up to 1994 for the purpose of assessing stock reclamation in the Gander River system. Following the methods of Ryan, Knoechel, and O'Connell (1994), we update the numerical relationship between adult small salmon (adults 63 cm) returning to the Salmon Brook fishway and the juveniles in two lakes at the headwaters of the Gander River system (as recruits) four years later. For the first time we examine the numerical relationship between those juveniles and the adults returning to the entire river system one year later (as determined from a counting fence and angler survey on the main stem of the river since 1989). Subsequently, we estimate juvenile abundance from 1995 to 1998 from the stock-recruit relationship and then project total river adult small salmon returns from 1995 to 1999 from the post-commercial fishery ratio of total returning adults to the EPA juvenile abundance one year earlier.

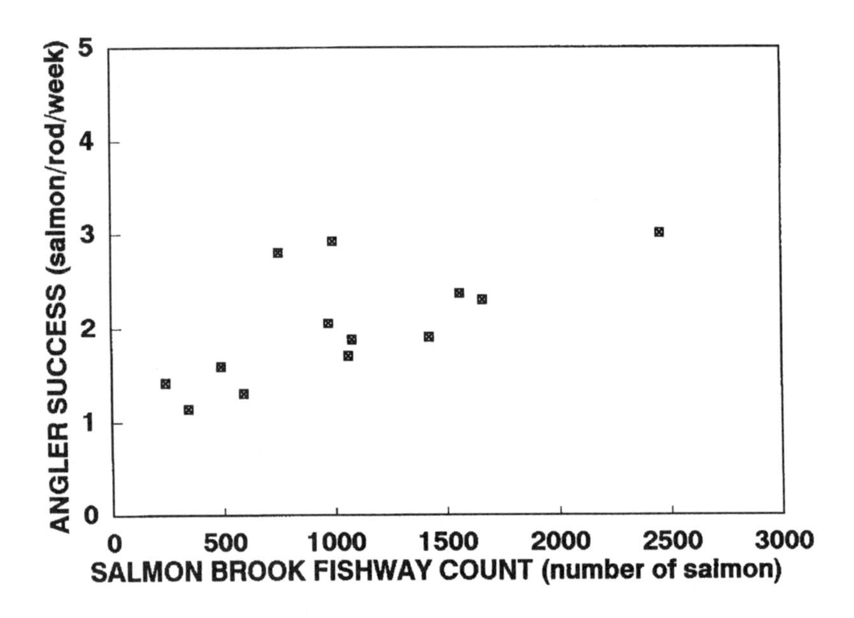

Figure 4. Proportionality between angler success for small salmon on the Gander River in its entirety and counts of small salmon at the Salmon Brook fishway, 1978-1991 (exclusive of partial fishway count of 1979). The trend between the variates is statistically significant: $r = 0.678$; $N = 13$; $p < 0.05$.

Juvenile Study Areas

Headwater and Spruce ponds are dilute (mean conductance 35 $uS \cdot cm^{-1}$), brown-water lakes within the Department of Fisheries and Oceans' Experimental Ponds Area (48° 19′N; 55° 28′W) at the headwaters of the Gander River system (Figure 5). Their physical and chemical characteristics approximate the average descriptors of water quality in insular Newfoundland (Ryan and Wakeham 1984). Headwater Pond (76.1 ha, maximum depth = 3.3 m, mean depth = 1.1 m) drains 3.5 km to the north into Spruce Pond (36.5 ha, maximum depth = 2.1 m, mean depth = 1.0 m), and the Spruce Pond outlet flows about 155 km northeast to the Atlantic Ocean. The closest known major concentration of salmon spawning substrate is about 12 km downstream of Spruce Pond (Ryan and Wakeham 1984). In addition to anadromous Atlantic salmon, other fishes present in these lakes are the brook trout *(Salvelinus fontinalis),* the American eel *(Anguilla rostrata),* and the threespine stickleback *(Gasterosteus aculeatus).* The history of ecological assessment in the Experimental Ponds Area has been reviewed by Ryan et al. (1994). Reviews of the population dynamics of salmon in the Experimental Ponds Area are available in Ryan (1993a, b) and references therein.

Adult Counting Sites

Detailed maps, analyses, and history of the Gander River fisheries and adult counting facilities are available in O'Connell and Ash (1992, 1993), Porter and O'Connell (1992), and references therein. Salmon Brook, a tributary, is downstream of Gander Lake on the main stem of the Gander River. Adult small salmon (adults < 63 cm) counted migrating through the fishway there represent 3.8-9.1% of those counted

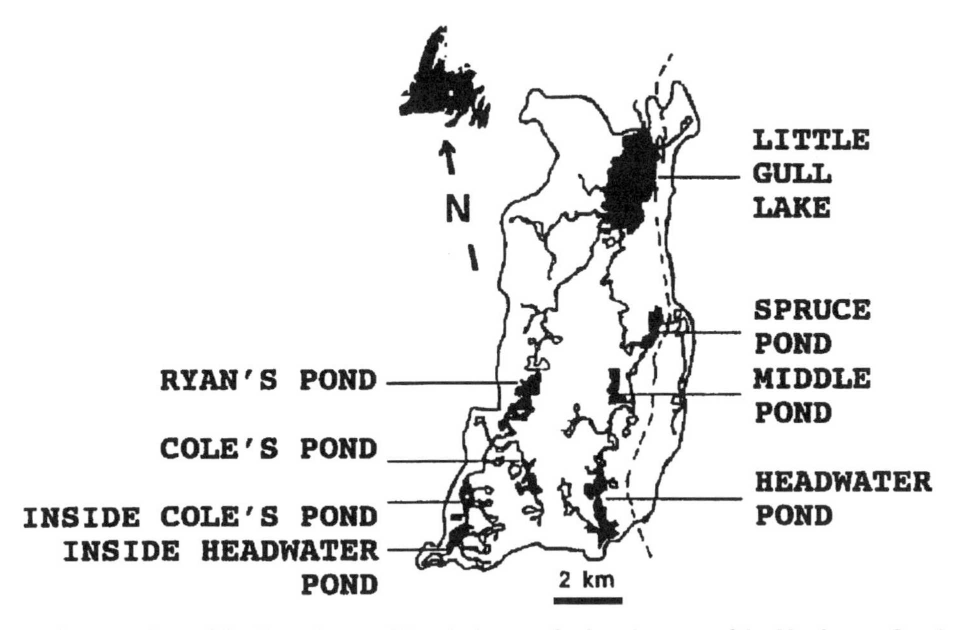

Figure 5. Watershed of the Experimental Ponds Area at the headwaters of the Northwest Gander River, central Newfoundland (inset). The dashed line through the east side of the watershed is the Bay D'Espoir highway.

at the Gander River counting fence on the main stem from 1989 to 1993 (O'Connell and Ash 1994).

An adult counting fence has been operated on the main stem of the Gander River since 1989, and total adult small salmon returns to the Gander River system have been calculated as the sum of the number of adults passing through the fence and the number angled downstream of the fence (O'Connell and Ash 1994). Small salmon have historically represented in excess of 90% of all adults escaping to the river (O'Connell and Ash 1994; Porter and O'Connell 1992).

Methods

Juvenile salmon abundance. Salmon were censused, concurrently with brook trout, in the spring and fall from 1978-1994 in Spruce Pond and from 1979-1994 in Headwater Pond using fyke nets and Schnabel multiple mark-recapture techniques as detailed by Ryan (1990). The study was terminated by management in 1988, but subsequently reinstated in 1989. Fish were captured in fyke nets, measured for length, marked with fin holes or clips, released, and recaptured for the computation of population size. Weights and scale samples have been routinely collected as documented by Ryan (1986a).

The age composition of the population during each census up to 1983 was calculated from the ages and lengths of subsampled fish, the lengths of released fish, the computed population size, and their relative proportions using age-length keys (Ricker 1975). Age-specific migrations to and from the lakes were calculated as the differences, by age group, between censuses. Thus, the number of salmon smolts migrating out of the lakes each year up to 1983 was calculated as the loss in numbers of salmon from each of the age-groups over the spring-to-fall period (Ryan 1986a). The calculated number of smolts in those years was related to the number of salmon present in the lakes in the spring of the year ($r=0.987$) by least-squares regression (Ryan 1986b). Accordingly, we used spring juvenile abundance here (Table 1) as a readily obtainable index of the smolt migration up to 1994.

Adult (small salmon) abundance. Small salmon counts at the Salmon Brook fishway in 1974 and from 1978 to 1994 were documented by O'Connell, Reddin and Ash (1995). Complete fishway counts were not obtained in 1979 but we used all available data from 1978 to 1994, including the partial count (Table 1).

Total adult small salmon returns to the Gander River system have been calculated as the number of adults passing through the counting fence on the main stem of the lower river plus the number angled downstream of the fence (O'Connell and Ash 1994). We have used all return counts since installation of the fence in 1989 as updated to 1994 by O'Connell, Reddin, and Ash (1995).

Juvenile-adult relationships. Comparisons between the number of juvenile salmon in the study lakes in spring and adult returns were made with linear regression analyses. In order to examine adults as a predictor of subsequent juvenile abundance (stock-recruit), we compared counts of salmon at the Salmon Brook fishway with juvenile salmon abundance four years later. The four year time delay accounted for the emergence of young the year following the count at the fishway plus an average

Table 1. Spring Atlantic salmon juvenile population sizes in the EPA projected to 1998, Salmon Brook fishway small salmon (<63 cm) counts, and Gander River small salmon returns projected to 1999

Year of Census	Spruce and Headwater ponds total Atlantic salmon juveniles (no. calculated '95 - '98)	Salmon Brook fishway count (no.) (yr N)* (partial count - '79, adjusted count -'90)	Gander River total small salmon returns (yr N) * (no. calculated '95 - '99 with 95% confidence interval bracketed)
1978		755	
1979	4822	404	
1980	3463	997	
1981	2393	2459	
1982	3077	1425	
1983	1603	978	
1984	3226	1081	
1985	3175	1663	
1986	4474	1064	
1987	3199	493	
1988		1562	
1989	4925	596	7743
1990	3642	345	7740
1991	2362	245	6745
1992	3069	1168	18179
1993	2470	1560	26205
1994	2370	963	18080
1995	1903		18605 (12231-24978)
1996	3543		14942 (3390-27687)
1997	4239		27810 (15053-42310)
1998	3179		33276 (18479-49859)
1999			24952 (12805-38836)

*From O'Connell, Reddin, and Ash: DFO Atlantic Fisheries Research Document 95/ 123.

age of three years of lake juveniles in the spring (Ryan 1986a). In order to examine juveniles as a predictor of adult abundance, juvenile abundance was compared to data on adult returns in the following year to reflect the predominant (94%) one year residence of adults at sea (see O'Connell and Ash 1994). Confidence limits (95%) about projected adult returns were obtained from the *t* distribution (1995 projection) and 20,000 Monte Carlo realizations (1996-1999 projections).

Detailed documentation of all data employed in this report has been presented by Ryan, Knoechel, O'Connell, Ash, and Warren (1995).

Results

Juveniles related to Salmon Brook adults four years earlier. The abundance of juvenile salmon in the Experimental Ponds Area has fluctuated during the period 1979 to 1994 with a maximum spring population of 4925 salmon in Spruce and Headwater ponds in 1989 (Figure 6). The pattern of seasonal change has been one of comparatively high spring abundance followed by a lower fall abundance after smoltification and the seaward migration.

The spring population size of Experimental Ponds Area juveniles demonstrated a strong stock-recruit relationship with adult small salmon returns monitored at the Salmon Brook fishway four years earlier, except for two notable outliers (fishway years 1981 and 1988) (Figure 7). With all data included, the relationship between juvenile abundance (Y) and adult returns four years earlier (X) was statistically significant:

Equation 1: $Y = 2167.370 + 0.909X$; $r = 0.619$; $N=12$; $p < 0.05$.

Anomalous environmental conditions, in the form of extreme regional flooding, are known to have occurred in January of 1983. In that month, precipitation at Bay D'Espoir was 325% of normal and 238.5 mm of rain were recorded on January 12-13, 1983. Precipitation on those two dates represented 77% of the total for the month (Environment Canada 1983). Travel by road to the Experimental Ponds Area at that time was not possible due to road washouts.

It was hypothesized that the January 1983 flooding may have been associated with an atypically high mortality of underyearling salmon during the winter of 1983, thereby resulting in a much lower than expected number of juveniles in the ponds in 1985. Additionally, the data point corresponding to the 1981 fishway count obviously was atypical within the 12-year trend. The data point was well outside the range of data we wished to use in our projections. Accordingly, we deleted fishway year 1981 from the regression for predictive purposes.

We are not aware of any unusual environmental circumstances associated with fishway year 1988 and we retained the associated data point for predictive purposes.

With the 1981 salmon count at the fishway (corresponding to juvenile year 1985) deleted from the stock-recruit curve (Figure 7), 74% of the variation in juvenile abundance (Y) was accounted for by adult returns four years earlier (X):

Equation 2: $Y = 1468.360 + 1.776X$; $r = 0.859$; $N=11$; $p < 0.01$.

Gander River adults related to juveniles one year earlier. Examination of the relationship between the spring estimate of abundance of Experimental Ponds Area juveniles and the total adult small salmon returns to the Gander River system in the subsequent year revealed a distinct separation of data points corresponding to the periods before and after the closure of the commercial fishery (Figure 8). The mean survival ratio index (ratio of adult returns to EPA juveniles in the previous year) of 1.71 for the two years prior to the closure of the commercial fishery in 1992, increased to a mean of 7.85 for the three years after ($t = 9.37$; $p < 0.01$) (Table 2). This difference was indicative of a more than fourfold increase in marine survival rates.

Projected juvenile abundance to 1998 and adult abundance to 1999. The regression equation 2 of Figure 7 provided calculated values of spring juvenile abundance in the Experimental Ponds Area for 1995-1998 (Figure 9). Use of those calculated juvenile

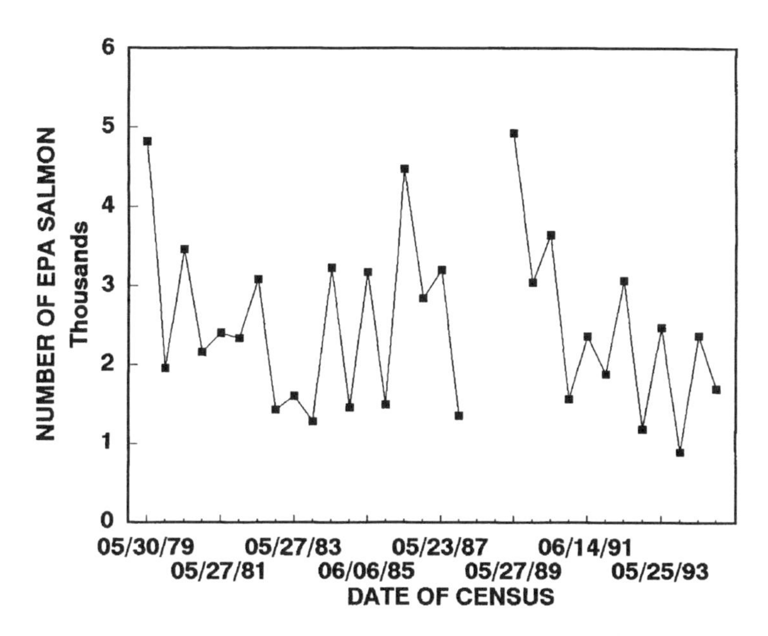

Figure 6. Schnabel population estimates of Experimental Ponds Area (EPA) juvenile salmon (Headwater and Spruce ponds combined) in the spring and fall, 1979-1994.

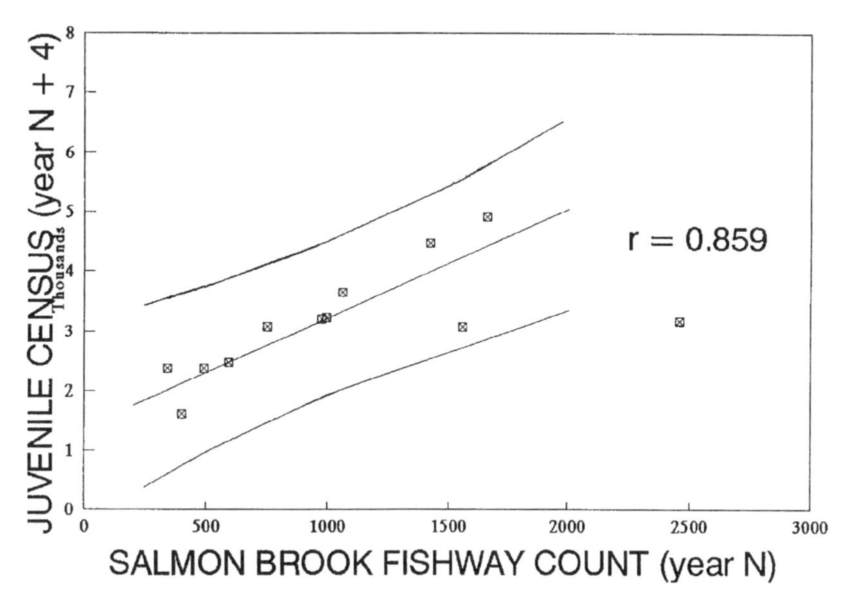

Figure 7. Stock-recruit relationship for the Gander River system based upon counts of small salmon (<63 cm) at the Salmon Brook fishway and the spring census of juveniles in the Experimental Ponds Area four years later. The two obvious outliers are fishway years 1981 (far right) and 1988. The regression equation with the 1981 fishway data removed is: Equation 2: $Y = 1468.360 + 1.776X$; $r = 0.859$; $N=11$; $p < 0.01$. The 95% confidence belt about the regression is shown.

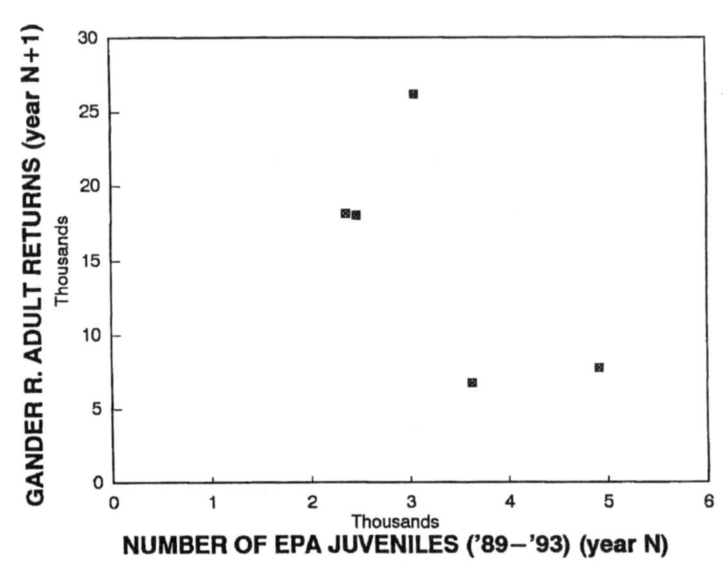

Figure 8. Decreased marine mortality of Gander River Atlantic salmon associated with the closure of the commercial salmon fishery in 1992 as indicated by total Gander River small salmon returns and the spring census of juveniles in the Experimental Ponds Area (EPA) one year earlier. The two data points at the lower right represent adult data from 1990 (far right) and 1991, while the upper three points correspond to the period of no commercial fishery.

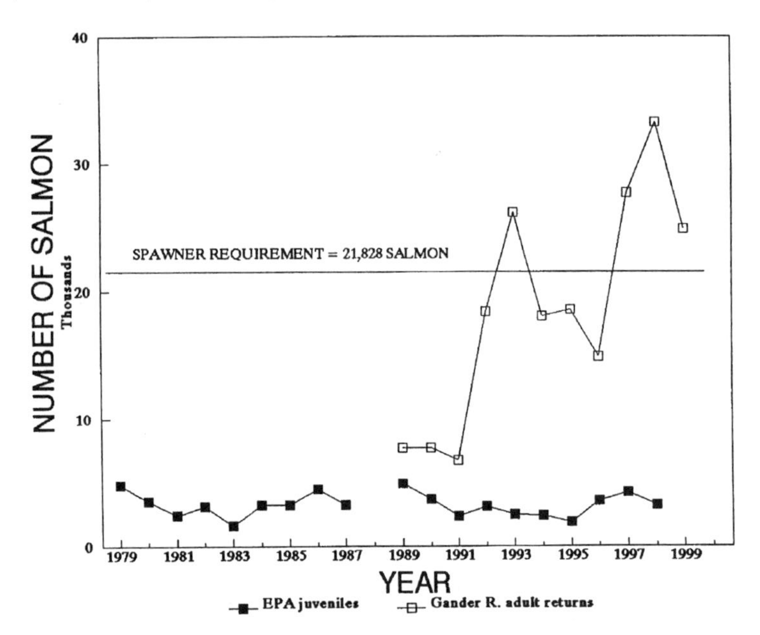

Figure 9. Experimental Ponds Area (EPA) spring juvenile salmon abundance to 1998 and projected Gander River small salmon returns to 1999. Juvenile numbers for 1995-1998 are calculated from equation 2 of Figure 7 while data from other years are measured values. Adult returns to 1994 are actual values while returns for 1995-1999 are calculated from the estimated juvenile numbers (from equation 2), and the mean ratio of adults to juveniles in the previous year after the closure of the commercial fishery. Presented for comparison is the estimated target spawning requirement for the Gander River.

Table 2. Ratios of Gander River total adult returns to EPA juveniles in the previous year with survival ratio indices before and after the closure of the commercial fishery in 1992			
Year of Census	**Spruce and Headwater ponds total Atlantic salmon juveniles (yr N)**	**Gander River total small salmon returns (yr N+1)**	**Survival ratio index (adults/juveniles)**
1989	4925	7740	1.57
1990	3642	6745	1.85
1991	2362	18179	7.70
1992	3069	26205	8.54
1993	2470	18080	7.32
		Mean ratio	5.40
		Mean pre-closure ratio (S.E.)	1.71 (0.198)
		Mean post-closure ratio (S.E.)	7.85 (0.625)

abundance values and the mean post-commercial fishery ratio of adults to juveniles in the previous year provided a first projection of Gander River small salmon returns to 1999 (Table 1, Figure 9).

Calculations indicate a high level of variability in future adult returns, but our projections do indicate more than a quadrupling of pre-moratorium returns to the Gander River by 1998. Returns in that year are predicted to be 33,276 adult small salmon.

Discussion

Validity of juvenile census data. The validity of census results for stock assessments has been verified previously by comparisons of calculated frequencies of marked fish in the lakes on the next-to-last sampling days with observed frequencies in the final census samples and by the relationships between catch per unit effort and census results (Ryan 1990). However, census data do not provide precise point estimates due to the fact that census assumptions of no emigration or immigration are not completely satisfied (Ryan 1990). This shortcoming has been minimized through empirical analysis and design optimization (Knoechel and Ryan 1994).

Accordingly, census results have provided a practical way of monitoring the juvenile stock. Additionally, it is apparent from the internal consistencies in the results herein (i.e. correspondence of juvenile densities to counts at the fishway and adult returns) that valuable inferences can be drawn concerning the status of the Gander River salmon.

Gander River stock-recruit curve. It is evident from Figure 7 that the abundance of adult salmon returning to the Gander River to spawn has been the major determinant of the subsequent abundance of juveniles. The outliers in the stock-recruit curve (Figure 7) likely resulted from one or more unusual environmental factors during the four-year interval between adult return and the subsequent assessment of juveniles, thereby resulting in a dramatic variation in year-class strength. It seems probable that flooding of the river system in January of 1983 was the cause of the reduced strength of the 1982 year class. Young salmon produced by the 1981 adult run would have been underyearling fish during the flood of January 1983.

Generally, it has been observed that salmon parr survival is related positively to stream discharge in both summer and winter. However, mortality rates of parr tend to be highest among underyearlings and, where suitable habitats are lacking, survival over winter is expected to be poor (Symons 1979; Gibson 1993). Young salmon may obtain higher survival rates by entering lakes but in insular Newfoundland the first lakeward migrations of salmon from streams typically occur after the first year of life (Ryan 1993b). Underyearling salmon in the stream environment during January of 1983 would have been subjected to displacement due to flood conditions and physical habitat disruption due to ice scour.

It appears that adult counts at the Salmon Brook fishway will serve as an indication of adult escapement to the Gander River but, since that fishway is located on a tributary well downstream on the river system, it may not provide an indication of recovery of the Gander stock in its entirety. The Salmon Brook tributary could be adequately seeded in the future while the remainder of the Gander River system remained below carrying capacity. Should this occur, it should become apparent in the form of radical deviations from the existing stock-recruit curve (Figure 7). We expect that, if the Salmon Brook fishway counts continue to be a representative measure of adult escapement, that the stock-recruit curve will serve as a means of evaluating angling management strategy and the attainment of maximum spawner capacity.

Impact of the fishery closure. The distinct separation of Figure 8 data points which correspond to the periods before and after the closure of the commercial fishery indicates a dramatic increase in juvenile to adult survival after the fishery closure. These results provide the first indication of the expected substantial increase in the marine survival of Gander River salmon following closure of the commercial fishery in 1992.

Projected time of stock recovery. The target spawning requirement for the Gander River, or the number of spawning salmon required for sustained production, has been estimated as 21,828 small salmon (O'Connell and Dempson 1991). From 1989 to 1991, low numbers of salmon returning to spawn in the Gander River system resulted in an egg deposition of about 35% of that required for the river (Porter and O'Connell 1992; O'Connell and Ash 1993). Results of the present study suggest that the number of salmon returning to the Gander River to spawn will not exceed the estimated spawning requirement in 1995 or 1996. Our projections indicate that the estimated spawning requirement should be exceeded from 1997-1999.

Large fluctuations in the number of adult returns, such as those evident from 1993 to projected 1999 levels, likely will persist due to variation in juvenile year class strength. These fluctuations will be evident in advance during future juvenile assessments and permit updating of potential harvest levels as stocks continue to recover. At this time, projected stock levels, by themselves, should not be used as a basis for stock management due to the fact that only three years of post-commercial fishery data are available. Substantial departures of freshwater and marine survival rates from recent levels may result in deviations from projected adult returns. Accordingly, the projections should, without corroboration, only be considered as first estimates and as a documented test of the methods employed for comparison with subsequent assessments.

Authors' note: At the time of the dismantling of the Gander River counting fence on September 4, 1995, the observed Gander River total small salmon returns for the year numbered 22,002 fish. The difference of 3,397 salmon between the observed returns and our projected returns for 1995 represents 15% of the actual 1995 count, indicative of a slightly faster than anticipated stock recovery.

Acknowledgments

A large number of Fisheries and Oceans staff and Memorial University students have assisted with data collections in the Experimental Ponds Area over the years. We especially appreciate the assistance of C.E. Campbell, K. Clarke, L.J. Cole, D.P. Riche, and D. Wakeham. Graduate students of Memorial University assisted with current data collections. The Newfoundland Region Atlantic Salmon Stock Assessment Committee reviewed the manuscript.

References

Environment Canada. 1983. Monthly record. Meteorological observations in Canada January 1983. Atmospheric Environment Service. Microfiche 1983 /01, Ref. 015.

Gibson, R.J. 1993. The Atlantic salmon in freshwater: spawning, rearing and production. *Reviews in Fish Biology and Fisheries* 3: 39-73.

Knoechel, R., and P.M. Ryan. 1994. Optimization of fish census design: an empirical approach based on long-term Schnabel estimates for brook trout *(Salvelinus fontinalis)* populations in Newfoundland lakes. *Verh. Internat. Verein. Limnol.* 25: 2074-2079.

Murray, A.R. and T.J. Harmon. 1969. A preliminary consideration of the factors affecting the productivity of Newfoundland streams. *Fish. Res. Board Can. Tech. Rep.* 130: vi + 405 p.

O'Connell, M.F. and E.G.M. Ash. 1992. Status of Atlantic salmon (*Salmo salar* L.) in Gander River, Notre Dame Bay (SFA 4), Newfoundland, 1989-1991. *Canadian Atlantic Fisheries Scientific Advisory Committee Research Document* 92/25. 22 p.

O'Connell, M.F. and E.G.M. Ash. 1993. Status of Atlantic salmon (*Salmo salar* L.) in Gander River, Notre Dame Bay (SFA 4), Newfoundland, 1992. *DFO Atlantic Fisheries Research Document* 93/30. 15 p.

O'Connell, M.F. and E.G.M. Ash. 1994. Status of Atlantic salmon (*Salmo salar* L.) in Gander River, Notre Dame Bay (SFA 4), Newfoundland, 1993. *DFO Atlantic Fisheries Research Document* 94/50. 17 p.

O'Connell, M.F. and J.B. Dempson. 1991. Atlantic salmon (*Salmo salar* L.) target spawning requirements for rivers in Notre Dame Bay (SFA 4), St. Mary's Bay (SFA 9), and Placentia Bay (SFA 10), Newfoundland. *Canadian Atlantic Fisheries Scientific Advisory Committee Research Document* 91/18. 14 p.

O'Connell, M.F., D.G. Reddin, and E.G.M. Ash. 1995. Status of Atlantic salmon (*Salmo salar* L.) in Gander River, Notre Dame Bay (SFA 4), Newfoundland, 1994. *DFO Atlantic Fisheries Research Document* 95/123. 25 pp.

Porter, T.R. and M.F. O'Connell. 1992. Effects of changing fishing mortality on Atlantic salmon (*Salmo salar* L.) egg deposition in Gander River. *Canadian Atlantic Fisheries Scientific Advisory Committee Research Document* 92/32. 16 p.

Ricker, W.E. 1975. Computation and interpretation of biological statistics of fish populations. *Bull. Fish. Res. Board Can.* 191, 382 p.

Ryan, P.M. 1986a. Lake use by wild anadromous Atlantic salmon, *Salmo salar,* as an index of subsequent adult abundance. *Can. J. Fish. Aquat. Sci.* 43: 2-11.

Ryan, P.M. 1986b. Prediction of angler success in an Atlantic salmon, *Salmo salar,* fishery two fishing seasons in advance. *Can. J. Fish. Aquat. Sci.* 43: 2531-2534.

Ryan, P.M. 1990. Sizes, structures, and movements of brook trout and Atlantic salmon populations inferred from Schnabel mark-recapture studies in two Newfoundland lakes. *Amer. Fish. Soc. Symp.* 7:725-735.

Ryan, P.M. 1993a. An extrapolation of the potential emigration of Atlantic salmon *(Salmo salar)* smolts from Newfoundland lakes in the absence of brook trout *(Salvelinus fontinalis).* p. 203-207. In R.J. Gibson and R.E. Cutting (eds .) The production of juvenile Atlantic salmon in natural waters. *Can. Spec. Publ. Fish. Aquat. Sci.* 118: 262 p.

Ryan, P.M. 1993b. Natural lake use by juveniles: a review of the population dynamics of Atlantic salmon in Newfoundland lakes, p. 3-14. In Ryan, P.M., M.F. O'Connell, and V.A. Pepper. 1993. Report from the workshop on lake use by Atlantic salmon in Newfoundland, Canada. *Can. MS Rep. Fish. Aquat. Sci.* 2222: iv + 54 p.

Ryan, P.M., M.H. Colbo , R. Knoechel, K. Clarke, and A. Cook. 1994. Fifteen years of freshwater ecosystem monitoring at the Experimental Ponds Area, Newfoundland — history and highlights, p. 15-30. In C.A. Staicer, M.J. Duggan, and J.J. Kerekes (eds.). Kejimkujik Watershed Studies: Monitoring and research five years after "Kejimkujik '88." Workshop Proceedings, Kejimkujik National Park, Nova Scotia, October 20-21, 1993. *Environment Canada Atlantic Region Occasional Report* No. 3. 276 p.

Ryan, P.M., R. Knoechel, and M.F. O' Connell. 1994. Population sizes of juvenile Atlantic salmon (*Salmo salar* L.) in lakes of the Experimental Ponds Area as a measure of stock recovery in the Gander River, Newfoundland. *DFO Atlantic Fisheries Research Document* 94/13. 17 p.

Ryan, P.M., R. Knoechel, M.F. O'Connell, E.G.M. Ash, and W.G. Warren. 1995. Atlantic salmon (*Salmo salar* L.) stock recovery in the Gander River, Newfoundland with projections to 1999. *DFO Atlantic Fisheries Research Document* 95/95, 16 p.

Ryan, P.M., M.F. O'Connell, and V.A. Pepper. 1993. Report from the workshop on lake use by Atlantic salmon in Newfoundland, Canada. *Can. MS Rep. Fish. Aquat. Sci.* 2222: iv + 54 p.

Ryan, P.M., and D. Wakeham. 1984. An overview of the physical and chemical limnology of the Experimental Ponds Area, central Newfoundland, 1977-82. *Can. Tech. Rep. Fish. Aquat. Sci.* 1320: v + 54 p.

Symons, P.E.K. 1979. Estimated escapement of Atlantic salmon *(Salmo salar)* for maximum smolt production in rivers of different productivity. *J. Fish. Res. Board Can.* 36: 132-140.

Benefits Agricultural Producers Experience When Switching From Conventional Seeding Systems to Low Disturbance Seeding Systems

by Doug McKell and Blair McClinton
Saskatchewan Soil Conservation Association, Indian Head, Saskatchewan, Canada

Abstract. Recent research and farmer experience have demonstrated both economic and agronomic benefits with the use of Low Disturbance Seeding (LDS) systems on operating farms. These benefits include reduced operating costs, improvement in soil quality over time, increased productivity per unit area of cultivated land, improved seed infiltration, and higher carbon levels in the soil.

Introduction

Recent research and farmer experience have shown that use of a Low Disturbance Seeding (LDS) system in farm operations can result in both economic and agronomic benefits. This paper addresses these benefits.

Reduced Operating Costs

Operating costs in a Low Disturbance Seeding (LDS) system generally will be lower when compared with those of conventional seeding systems. This is a function of the number of tillage passes eliminated and replaced with LDS operations. The major savings will be in fuel, labour, and machinery depreciation. Most farmers adopting the LDS system have experienced savings of approximately 40% in fuel and labour (Branik Resources 1992). Although savings in depreciation are hard to quantify, farmers using LDS systems experience longer tractor life, fewer repairs, and, in some cases, lower overall capital investments in machinery. Research at Indian Head also has shown reduced operating costs with the zero-till system as compared to the conventional seeding system (Lafond et al. 1993).

Gray et al. (1994) used the whole-farm budget approach for a 1600-acre farm in central Saskatchewan. Using producers' suggestions for typical yields, operations and expenses, their model zero-till LDS system, with a 10% yield advantage, resulted in an ending net worth of $413,000 after five years of cropping. This was an advantage of $90,000 over the conventional seeding system or $18,000 per year ($11/acre).

Reduced Soil Erosion

Soil erosion is one the most serious threats to sustaining agricultural productivity in western Canada (Prairie Farm Rehabilitation Administration [PFRA] 1983). Erosion results from the action of wind and/or water on soils that are not adequately protected by vegetative cover. The benefits of maintaining crop residues to reduce erosion by wind or water are well known (PFRA 1983; Anonymous 1991). More residue is conserved when the number of tillage operations is reduced. LDS leaves behind up to 80% ground cover depending on the residue type and amount (e.g. wheat, barley, canola, lentil, field pea), and seeder design. This compares with 0-20% ground cover under conventional tillage systems. It has been suggested that no-till planting (LDS) can reduce erosion by 90% of what occurs on clean-tilled fields (Dickey et al. 1990). LDS has the potential either to eliminate or significantly reduce summerfallow acres, thereby reducing soil erosion.

Improved Soil Quality

Soil quality is related directly to soil organic matter levels. Soil fertility factors, like organic N, P, S levels, tend to increase with organic matter. Greer and Schoenau (1992) have found that management has a significant influence on the concentration of organic N and S in the 0-5 cm layer. Less mixing and dilution of the soil organic matter results in higher concentrations in the direct-seeded field than the conventional-tilled field. The mineralizable fraction of the total organic N and S also are higher in the direct seeded field than in the conventional-tilled field. This is consistent with a study by Selles et al. (1984), who has found that while there are no differences in N abundance by tillage system, the N in the zero-till plots is in more labile forms. In addition, Arshad and Coy (1993) reported improvements in aggregate stability.

The long-term crop rotation studies in western Canada have shown that soil quality improves when summerfallow is reduced (Campbell et al. 1990). Several researchers in western Canada have suggested that direct seeding has the potential to either eliminate or significantly reduce summerfallow frequency in all but the most arid areas of the Canadian prairies (Brandt et al. 1993; Lindwall et al. 1993; Lafond et al. 1992). This suggests that the combination of eliminating summerfallow and reducing tillage will lead to the greatest improvement in soil quality.

Increased Productivity

Several studies have measured differences in yield achieved with LDS techniques. Overall, the evidence suggests a yield advantage with zero-till systems. For example, in Saskatchewan, 5%-25% yield advantages have been reported at Melfort, up to 18% increases at Indian Head, and advantages of 11.4% for wheat and 5.7% for oilseeds at Scott (Branik Resources 1993). These yield increases with LDS have been attributed to improved moisture conservation from improved snow trapping, infiltration, and reduced evaporation (Lafond et al. 1992). Additional improvements in productivity may be possible as the soil quality improves.

Improved Infiltration

Anecdotal evidence from long-term direct seeding farmers and researchers suggests that infiltration rates improve over time under direct seeding conditions (Coutts and Smith 1991; SSCA 1992; S.A. Brandt, personal communication). They claim that surface ponding after rainfalls is reduced significantly on direct seeded fields. This reduces crop losses from flooding and allows for more timely field operations. This improvement in infiltration is supported by U.S. research (Maulé 1993; Hatfield et al. 1993).

Research on water infiltration as it applies to conservation tillage systems in western Canada is limited. Preliminary research by Maulé and Reed (1993) has indicated that infiltration rates tend to improve the longer a field is in no-till (LDS). They have found that infiltration rates increase with soil organic matter levels. There are no significant differences between the no-till fields and a continuously cropped conventional-tilled field with high organic matter levels. However, the no-till fields have shown significantly greater saturated hydraulic conductivities than have conventional-tilled fields.

Increased Soil Carbon

LDS increases soil carbon in two ways: (1) LDS increases biomass production by reducing the frequency of summerfallow and increasing crop yields, and (2) it reduces mineralization processes by minimizing mixing crop residues into the soil. Experimental and simulated results on the impacts of changing agricultural land use on atmospheric C show that the reduction of fallow area, adequate fertilization, and elimination of soil erosion through the use of zero tillage (LDS), have the potential to add approximately 0.4% C to the 0-30 cm soil carbon pool under cereals and 2.0% under hay (Dumansky et al. 1995). It has been suggested that under these conservation management techniques soil carbon will increase over the first 15-25 years before reaching a steady state (Dumansky et al. 1995).

Results from research performed at Swift Current, Saskatchewan were published recently. In "Managing Soils to Store Carbon More Effectively," Campbell et al. (1994) showed that soil carbon increases under an LDS system at rates of about 0.4-0.5 tonnes/hectare/year. This result occurred over a 12-year period from 1982-1994 when the conventional tillage based half-and-half seeding system was replaced with a continuous wheat rotation, no-tillage system. After years of no-tillage, they reverted to mechanical tillage and found the organic matter and the resulting soil carbon to decline slowly. This happened despite the retention of a continuous cropping rotation. Projects which increase the adoption of practices which increase soil carbon, like LDS, can have a major impact on partially offsetting greenhouse gas emissions.

References

Anonymous. 1991. Tillage practices that reduce erosion. Soil Works. Canada-Saskatchewan Agreement on Soil Conservation. 4 p.

Arshad, M.A. and G.R. Coy. 1993. Tillage and cropping systems for soil conservation and sustained crop production in the Peace River region. Final Report, Canada-Alberta Agreement on Research and Technology Transfer (CARTT). 54 p.

Brandt, S.A., R.P. Zentner, C.A. Campbell, and V.O. Biederbeck. 1993. Crop rotations for direct seeding systems. *Proc. direct seeding: making it work in the drier soil zones*. Saskatchewan Soil Conservation Association. Moose Jaw, SK. pp. 8-18.

Branik Resources. 1992. Survey of producers using the conserva pak zero till seeding system. Published by Agricultural and Bioresource Engineering Department, University of Saskatchewan, Saskatoon, SK.

Branik Resources. 1993. Economics of zero tillage. Soil Works. Canada-Saskatchewan Agreement on Soil Conservation. 24 p.

Campbell, C.A., B.G. McConkey, R.P. Zentner, and F. Selles. 1994. Managing soils to store carbon more effectively. Research Newsletter, No. 7. Swift Current Research Station, Agriculture and Agri-food Canada. Nov. 18, 1994.

Campbell, C.A., R.P. Zentner, H.H. Janzen, and K.E. Bowren. 1990. Crop rotation studies on the Canadian Prairies. Agriculture Canada publication 1841/E. 133 p.

Coutts, G.R. and R.K. Smith. 1991. Zero tillage production manual. Manitoba-North Dakota Zero Tillage Farmers Association. 43 p.

Dickey, E. and D. Shelton. 1990. Residue, tillage and erosion. *Proc. Great Plains Conservation Tillage Symposium*. Bismarck, ND. pp. 73-83.

Dumansky, J., C. Tarnocai, C. Monreal, R.L. Desjardins, E.G. Gregorich, and C.A. Campbell. 1995. Possibilities for future carbon sequestration in Canadian agriculture in relation to land use changes. Global Change Biology. Submitted March 1995.

Gray, R., J. Taylor, and W. Brown. 1994. Economic factors contributing to the adoption of reduced tillage technologies. In D. Derksen, G. Lafond, J. Moen, eds. *Proc. implications of crop residue management and conservation tillage*. Regina, SK. pp. 116-124.

Greer, K.J. and J.J. Schoenau. 1992. Soil organic matter content and nutrient turnover in Thin Black Oxbow soils after intensive conservation management. *Proc. soils and crops workshop*. Saskatoon, SK. pp. 167-173.

Hatfield, J.L. and J.H. Prueger. 1993. Soil tilth effects of surface residue management systems. *Proc. Manitoba-North Dakota Zero Tillage Farmers Association workshop*. Brandon, MB. pp. 13-30.

Lafond, G.P., H. Loeppky, and D.A. Derksen. 1992. The effects of tillage systems and crop rotations on soil water conservation, seedling establishment and crop yield. *Can. J. Plant Sci.* 72: 103-115.

Lafond, G.P., R.P. Zentner, R. Geremia, and D.A. Derksen. 1993. The effects of tillage systems on the economic performance of spring wheat, winter wheat, flax and field pea production in east-central Saskatchewan. *Can. J. Plant Sci.* 73: 47-54.

Lindwall, C.W. and F.J. Larney. 1993. Why direct seeding will work and is profitable. *Proc. direct seeding: making it work in the drier soil zones*. Saskatchewan Soil Conservation Association. Moose Jaw, SK. p. 5.

Maulé, C.P. and W.B. Reed. 1993. Infiltration under no-till and conventional tillage systems in Saskatchewan. *Can. Agric. Engineering* 35(3): 165-173.

Maulé, C.P. 1993. Improving water use efficiency. *Proc. direct seeding: making it work in the drier soil zones*. Saskatchewan Soil Conservation Association. Moose Jaw, SK. p. 19-28.

PFRA. 1983. Land degradation and soil conservation issues on the Canadian Prairies. Prairie Farm Rehabilitation Administration (PFRA). 326 p.

SSCA. 1992. How to: direct seed cereals and oilseeds. Soil conservation: Video Guide II. Saskatchewan Soil Conservation Association. 26:20 min.

Selles, F., R.E. Karamanos, and K.E. Bowren. 1984. Changes in natural ^{15}N abundance of soils associated with tillage practices. *Can. J. Soil Sci.* 64: 345-354.

Subsurface Drainage for Soil Salinity Reclamation of an Irrigated Soil at the Saskatchewan Irrigation Development Centre

by T.J. Hogg and L.C. Tollefson

Agronomist and Manager, Saskatchewan Irrigation Development Centre, Outlook

Abstract. Subsurface drainage for soil salinity reclamation commenced on a 9.0 hectare field at the Saskatchewan Irrigation Development Centre (SIDC) in 1986. A linear sprinkler irrigation system was used to apply leaching water in the fall of each year after harvest, beginning in 1988. The effects of this leaching on drainage outflow and effluent quality were monitored. Results indicated large quantities of salt were removed with the leaching water each year. Changes in soil salinity were monitored with an EM38 non-contacting terrain conductivity meter. In addition, soil samples were collected at permanently located benchmark sites. Results based on the top 0.75 metres of the soil profile indicated a reduction in the area of moderately plus severely saline soils from 62% prior to drainage installation in 1986 to only 3% in 1989, with little change thereafter. Dramatic improvement in grain yield occurred since drainage installation and fall leaching. In 1995, a crop of dry beans which are considered to be salt sensitive, was produced on this land. This indicated that reclamation of this area was achieved.

Introduction

Irrigation is necessary for successful crop production in many regions of the world. However, improper water management can result in soil salinity and water logging. Reclamation is achieved by controlling excess water inputs and providing adequate drainage to lower the water table and leach the accumulated salts (Hoffman 1985). The effectiveness of leaching varies among soils. The degree of salt removal depends on the quantity of water that passes through the soil (Reeve and Fireman 1967). The amount of leaching water and the time required to reclaim a soil depends on several factors including the depth of soil to be reclaimed, the initial salinity level, the type of salts present and soil characteristics such as texture, structure, infiltration and permeability (Rhoades 1982a). Generally, one depth of leaching water will decrease the salt concentration of an equivalent depth of soil by 70-80% (Bole 1986; Harker and Mikalson 1990; Jury et al. 1979; Rapp 1968; Reeve et al. 1955; van Schaik and Milne 1962).

The Saskatchewan Irrigation Development Centre (SIDC), located at Outlook,

Saskatchewan, has undertaken irrigated crop production since 1949. A rise in the water table has increased soil salinity (Jones and Lebedin 1986).

These salinity problems were caused by poor surface water control and inadequate internal soil drainage. Reclamation requirements were determined and suggested to include subsurface tile drainage installation (Jensen and Wright 1986). In 1986, subsurface drainage was installed to lower the water table and provide a means for leaching out the excess soluble salts. In addition, the water delivery and surface drainage systems were improved. Flood irrigation was replaced with a sprinkler irrigation system, allowing greater precision in water application. Improvements to the surface drainage provided better control of surface runoff and prevented surface water ponding.

This paper reports on the effectiveness of subsurface drainage and subsequent leaching on the reclamation of saline and waterlogged irrigated land at SIDC.

Methodology

Site characteristics. The drainage site was located on a 9.0 ha field in the southwest corner of SIDC (SW15-29-08-W3 - Field 11). The area consists of both well-drained calcareous Bradwell soils and imperfectly drained saline Bradwell soils with a loam to silty loam surface texture (Stushnoff and Acton 1987). The subsoils consist of fine textured lacustrine material overlying glacial till. The lower permeable subsoil forms a barrier to the deep percolation of water.

Drainage installation. The drainage system was designed using the Hooghoudt equation (Wither and Vipond 1983):

$$W^2 = \frac{4}{R}(2deK_2h + K_1h^2)$$

W = spacing between drains (m)

R = drainage rate (m/day)

de = effective depth of flow below drains (m)

K_2 = saturated hydraulic conductivity below the drains (m/day)

h = height of the water table at the mid spacing between the subsurface drains measured above the centre line of the drains (m)

K_1 = saturated hydraulic conductivity above the drains (m/day)

Values used in the equation to estimate drain spacing were based on previous drainage experience (R = 0.001 m/day; de = 1 m; h = 0.7 m) (Jensen and Wright 1986) and measured values from hydrogeological studies of the site (K_2 = 0.009 m/day; K_1 = 0.864 m/day) (Jones and Lebedin 1986).

Polypropylene drainage pipe fitted with a polyester filter sock was installed at 15 m and 30 m spacings, diagonally across the field parallel to the surface drain, using a laser trencher (Figure 1). Laterals (100 mm) were buried on a 1.2 to 2.2 m gradient and were connected to main conduits (150 mm) buried on a 1.6 to 2.6 m gradient. The main conduits drain into an effluent collector located in the southwest corner of the field. Drainage effluent is carried by pipeline (300 mm) and discharged to the South Saskatchewan River.

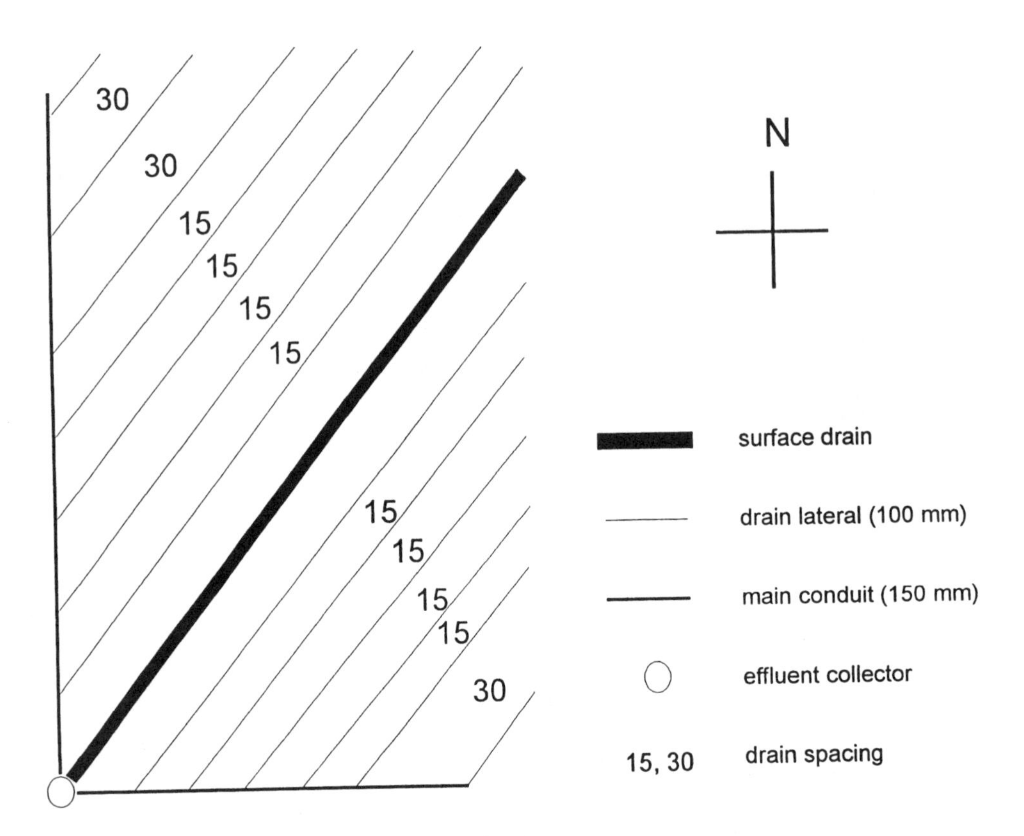

Figure 1. Subsurface drainage layout on Field 11.

Leaching. Since 1988, large applications of irrigation water were applied in the fall after harvest to facilitate leaching of salts. Drainage flow rates were determined by timing the collection of a measured volume of effluent. Salt content in the drainage effluent was determined from electrical conductivity and chemical analyses.

Soil monitoring. A monitoring program was established to determine the change in soil salinity after leaching water application. Bulk soil electrical conductivity (ECa) measurements were taken using an EM38 non-contacting terrain conductivity meter. This instrument measures ECa to a depth of approximately 1.5 m in the vertical dipole orientation and 0.75 m in the horizontal dipole orientation (McNeill 1986). The meter response is due mainly to salinity. EM38-ECa readings also are affected by moisture, temperature, soil texture, and soil mineralogy. The influences due to moisture and temperature were minimized by taking readings at the same time each year. A permanent grid (15 m x 15 m) was established in 1986, and EM38 readings were taken in October on a yearly basis. Soil samples were collected at specific grid points and analyzed for electrical conductivity (ECe) and major cation and anion content on the saturated paste extract (Rhoades 1982b).

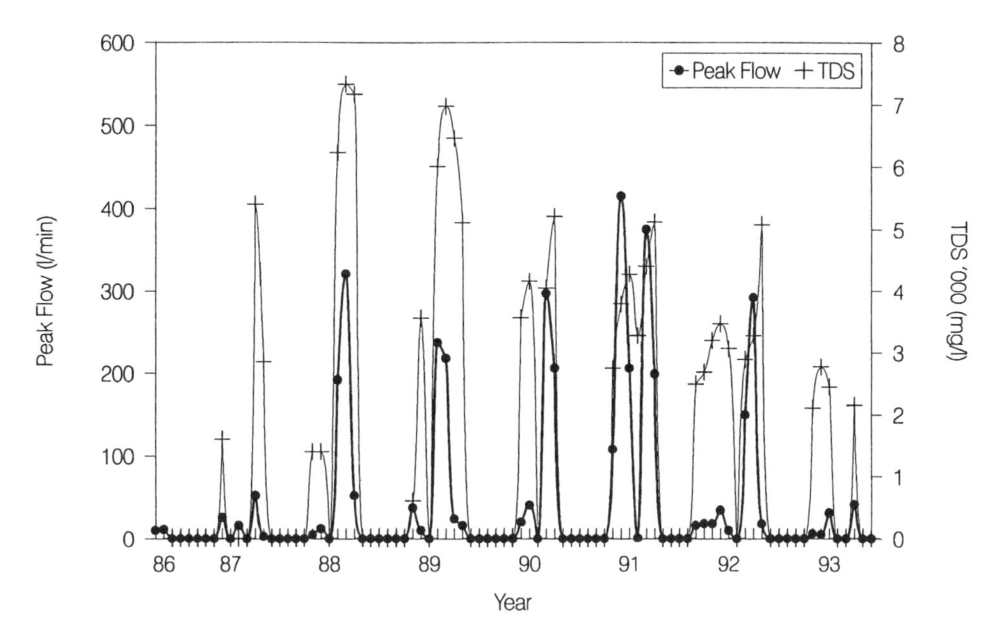

Figure 2. Monthly peak drain flow rates and TDS on Field 11 after subsurface drainage installation.

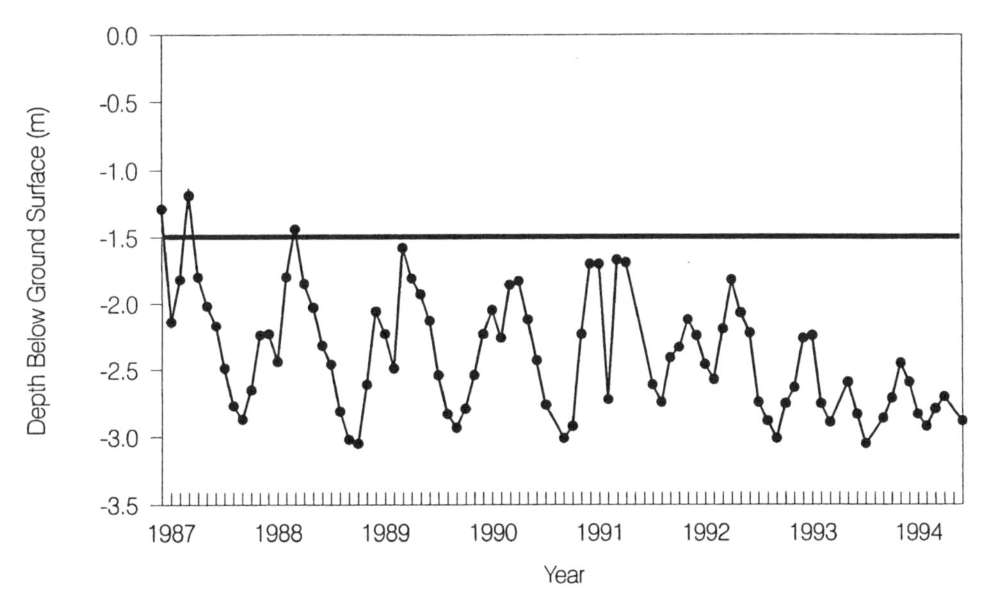

Figure 3. Mean monthly water table levels on Field 11 after subsurface drainage installation (1987-1994).

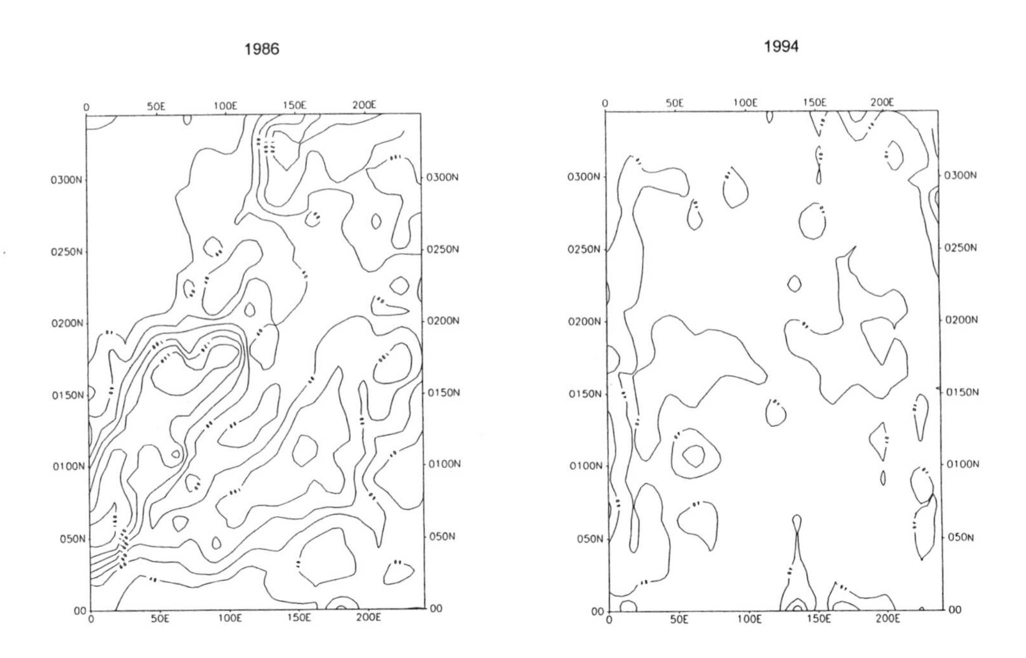

Figure 4. EM38 horizontal bulk soil salinity (ECa) contour map of Field 11 for 1986 and 1994.

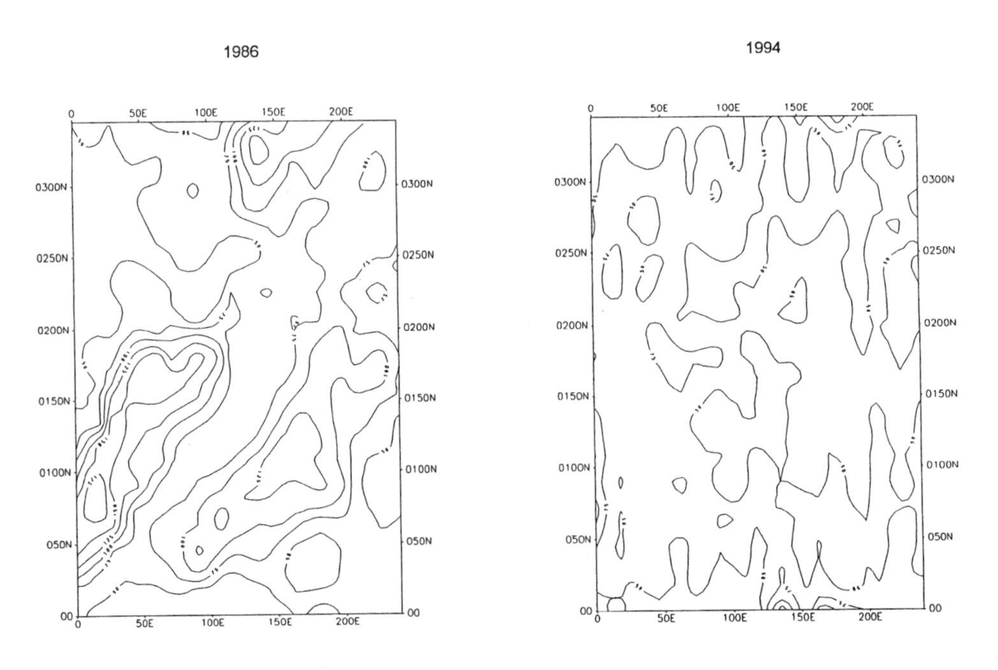

Figure 5. EM38 vertical bulk soil salinity (ECa) contour map of Field 11 for 1986 and 1994.

Results

Leaching. The effect of leaching water on drainage water flow rate and effluent content is illustrated in Figure 2. Monthly peak drain flow rates during the growing season ranged from 0 to 400 l/min in 1991 and were generally influenced by crop growth and rainfall patterns. During the fall leaching period, monthly peak drain flow rates increased with application of leaching water and ranged between 200-400 l/min. The corresponding total dissolved solids (TDS) of the drainage effluent at these flow rates, estimated from effluent electrical conductivity: $TDS = 765.1\ EC^{1.087}$ (Chang et al. 1983), varied between 3000-7000 mg/l. At a flow rate of 200 l/min and a drainage effluent TDS of 3000 mg/l, approximately 36 kg of salt would be removed in one hour and approximately one tonne in a 24-hour period.

Changes in water table level on the drainage site are shown in Figure 3. The water table levels were maintained at or below the level of the drain laterals (1.5 m) after drainage installation similar to results obtained by Bennett et al. (1982). The water table rose during the growing season and peaked during the fall leaching period each year. After the fall leaching period the water table level declined until the commencement of the following irrigation season.

Soil monitoring. The EM38-ECa data collected each fall were gridded and contoured using "Geosoft," a geostatistical software package. Contour maps of the EM38-ECa readings indicated the location and severity of soil salinity present. The higher the ECa values the greater the salt content in the soil. EM38-ECa contour maps comparing 1986 and 1994 in both the horizontal (Figure 4) and vertical (Figure 5) dipole orientations indicate the changes in soil salinity that have occurred. A paired t-test conducted on the 402 grid ECa readings for 1986 and 1994 confirms the significant reduction in soil salinity during this period (Table 1).

The EM38-ECa data and ECe analyses collected at specific grid points were related by regression analysis (Harron and Tollefson 1989) to facilitate the interpretation of EM38-ECa readings into different soil salinity classes (Table 2). Geosoft then was used to calculate the areal extent of salinity at contour intervals determined to represent non (0-2 dS/m), slight (2-4 dS/m), moderate (4-8 dS/m), and severe (8-16 dS/m) salinity classes. The changes in salinity classes from 1986 to 1994 are indicated in Table 3. Results based on the ECa horizontal readings (0-0.75 m) indicate a reduction in the

Table 1. Paired t-test for EM38-ECa changes on Field 11

	Vertical	Horizontal
Mean ECa 1986	78	70
Mean ECa 1994	54	37
Mean Difference	24	33
t	13.85	20.23
df	401	401
t(0.01)	2.59	2.59

Table 2. Comparable EM38-ECa values for soil salinity classes			
Salinity Class	**ECe (dS/m@25C)**	**ECa Vertical**	**ECa Horizontal**
Non	0-2	0-50	0-40
Slight	2-4	50-72	40-58
Moderate	4-8	72-114	58-93
Severe	8-16	114-199	93-163

Table 3. Changes in soil salinity class from EM38-ECa horizontal readings (1986-1994)								
Salinity Class	**% Area**							
	1986	1988	1989	1990	1991	1992	1993	1994
Non (0-2 dS/m)	20	17	69	46	89	81	93	74
Slight (2-4 dS/m)	18	54	28	47	11	17	8	24
Moderate (4-8 dS/m)	34	26	3	8	0	2	0	2
Severe (8-16 dS/m)	28	3	0	0	0	0	0	0

Table 4. Paired t-test for soil ECe changes at 15 locations on Field 11 (1986-1992)						
Years Compared	**0-0.6 m**		**0.6-1.2 m**		**0-1.2m**	
	Mean ECe Difference (dS/m)	**t[1]**	**Mean ECe Difference (dS/m)**	**t**	**Mean ECe Difference (dS/m)**	t
1986-1988	5.7	7.15**	NA[3]		NA	
1986-1989	7.0	8.19**	4.1	4.00**	5.5	6.40**
1986-1990	6.9	7.20**	4.1	4.00**	5.5	6.00**
1986-1991	7.5	8.54**	4.5	4.07**	5.9	6.23**
1986-1992	7.2	8.53**	4.1	3.69**	5.6	5.97**
1988-1989	1.3	4.75**	NA		NA	
1988-1990	1.2	3.65**	NA		NA	
1988-1991	1.8	5.04**	NA		NA	
1988-1992	1.5	5.20**	NA		NA	
1989-1990	-0.1	-0.1NS[2]	0	-.58NS	0	-1.15NS
1989-1991	0.5	1.89NS	0.4	2.01NS	0.4	1.93NS
1989-1992	0.2	0.96NS	0	0.22NS	0.1	0.26NS
1990-1991	0.6	1.65NS	0.4	1.85NS	0.4	2.64*
1990-1992	0.3	1.17NS	0	0.68NS	0.1	1.74NS
1991-1992	-0.3	-2.37*	-0.4	-2.10NS	-0.3	-2.75*

[1]** t0.01, 14 df>2.98; *t0.05, 14df>2.14
[2]NS - not significant
[3]NA - no samples available

Table 5. Paired t-test for soil SAR changes at 15 locations on Field 11 (1986-1992)

Years Compared	0-0.6 m		0.6-1.2 m	
	Mean SAR Difference	t[1]	Mean SAR Difference	t
1986-1988	2.3	4.51**	NA[3]	
1986-1989	3.0	4.23**	0.4	0.60NS
1986-1990	3.0	4.11**	0.7	1.37NS
1986-1991	3.5	5.03**	1.1	1.12NS
1986-1992	3.1	4.33**	0.8	1.31NS
1988-1989	0.7	-0.98NS[2]	NA	
1988-1990	0.7	-0.97NS	NA	
1988-1991	1.2	1.26NS	NA	
1988-1992	0.8	-0.69NS	NA	
1989-1990	0	-0.29NS	0.3	0.81NS
1989-1991	0.5	2.86*	0.7	0.59NS
1989-1992	0.1	0.53NS	0.4	0.60NS
1990-1991	0.5	2.38*	0.4	0NS
1990-1992	0.1	0.99NS	0.1	0.17NS
1991-1992	-0.4	-2.17*	-0.3	0.21NS

[1]** $t_{0.01, 14df} > 2.98$; * $t_{0.05, 14df} > 2.14$
[2]NS - not significant
[3]NA - no samples available

area classified as moderately plus severely saline from 62% in 1986 to 29% in 1988 after the first fall leaching period. This was reduced to 3% in 1989 after the second fall leaching period, with little change thereafter.

Mean soil ECe, an indicator of soil salinity, over the study period is illustrated in Figure 6. A paired t-test comparing mean soil ECe for 15 permanent grid point sampling locations indicated significant reductions in soil ECe in the top 0.6 m of the soil profile after the first two leaching periods in the fall of 1988 and 1989 (Table 4). Accordingly, the soluble salt content of the soil has been leached past the 0.6 m depth. A significant reduction in salinity level also has occurred in the top 1.2 m of the soil profile.

Mean soil sodium adsorption ratios (SAR), calculated from cation content of the saturated paste extract: $SAR = Na/[(Ca+Mg)/2]^{1/2}$, concentrations expressed in meq/l (U.S. Salinity Laboratory Staff 1954), indicate changes similar to mean soil ECe changes (Figure 7). A paired t-test comparing mean soil SAR showed that the major changes occurred in the top 0.6 m of the soil profile after the first leaching period in 1988 (Table 5). There were no significant changes in mean soil SAR in the 0.6-1.2 m soil depth.

The effectiveness of leaching water application on the reduction in initial soil salt

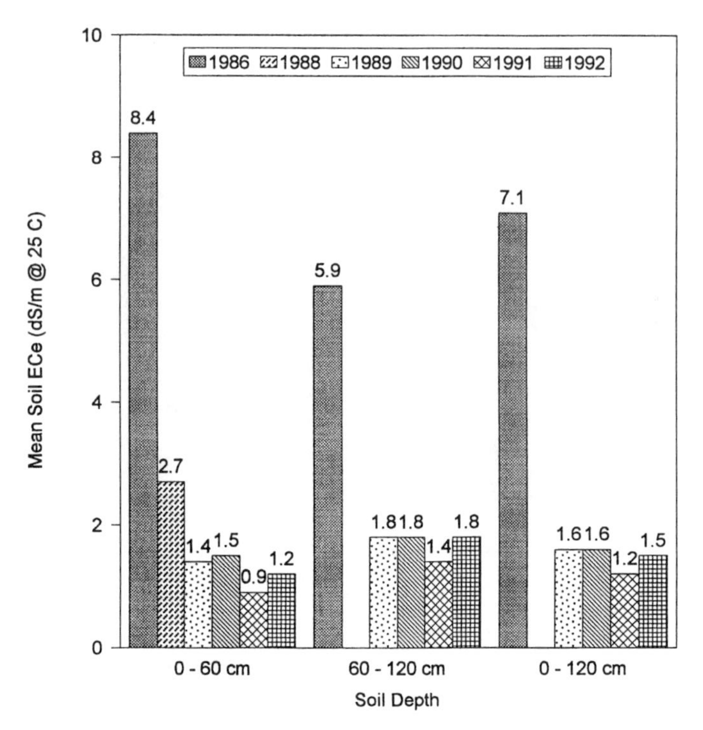

Figure 6. Mean soil ECe for 15 sites on Field 11.

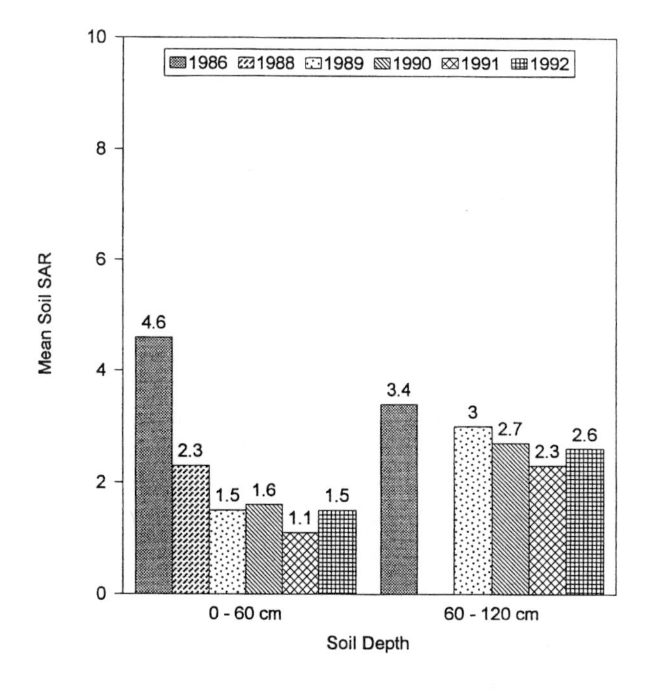

Figure 7. Mean soil SAR for 15 sites on Field 11.

Table 6. Total fall leaching water applied to Field 11

Year	Irrigation	Precipitation	Total Fall Leaching Water
		(mm)	
1988	475	43	518
1989	355	127	482
1990	270	20	290
1991	297	13	310
1992	222	15	237
1993	135	37	172
1994	50	5	55

Table 7. Yield of barley after subsurface drainage installation and leaching on Field 11

Year	Crop	Yield (kg/ha)
1988	Heartland Barley	3225
1989	Bonanza Barley	4470
1990	Duke Barley	6990
1991	Duke Barley	5900
1992	Duke Barley	6505
1993	Manley Barley	4838
1994	Duke Barley	5240

content can be obtained from the relationship developed by Reeve et al. (1955). In this relationship the reduction in initial soil salt content, C/Co, is expressed as a function of the depth of leaching water applied per unit depth of soil, Dw/Ds. In general, 50% of the initial salt present in the soil is removed when Dw/Ds = 0.5 and 80% when Dw/Ds = 1.0.

The total quantity of leaching water applied each fall after the crop was removed is indicated in Table 6. Approximately 500 mm of leaching water was applied to Field 11 by the fall of 1988 and about 1000 mm by 1989. This suggests that sufficient water had been applied to remove 50% of the salt to a depth of 100 cm after the first leaching period. By the fall of 1989, 80% of the salt to a depth of 100 cm should have been removed in the leaching water.

The changes in initial soil salt content, C/Co, calculated from the change in soil ECe expressed as a function of the depth of leaching water applied per unit depth of soil, Dw/Ds, is presented in Figure 8. The progression is similar to the relationship derived by Reeve et al. (1955). Other studies in western Canada have indicated the application of a depth of leaching water equal to the depth of soil reduced salinity by

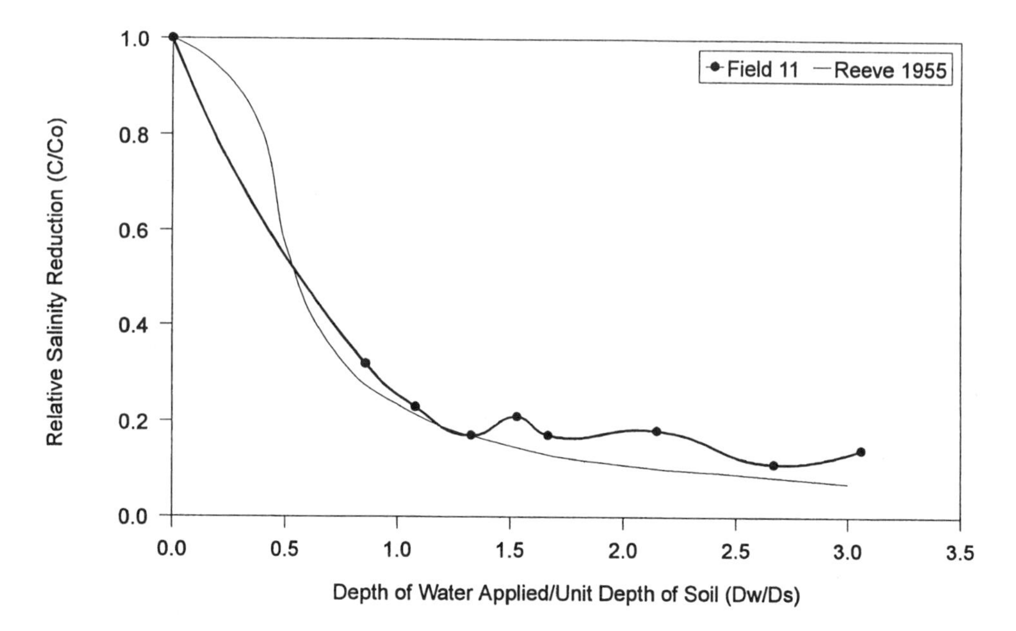

Figure 8. Relative salinity reduction versus leaching water application on Field 11 after subsurface drainage installation.

70% (Rapp 1968), 73% (van Schaik and Milne 1962), 75% (Harker and Mikalson 1990) and 76% (Bole 1986).

Crop response. Prior to installation of the drainage system, crops grown were limited to salt tolerant grass-legume mixtures with relatively poor production. The grass-legume cover was broken in the fall of 1987 and barley was seeded in 1988. The crop grew primarily above the drains where soil mixing had occurred. Marked improvement in barley yield occurred over the period between 1988 and 1994 (Table 7).

The growth of salt-sensitive crops such as legumes would give a better indication of the degree of reclamation of soil salinity in this field. Bean (*Phaseolus vulgaris*), a salt-sensitive crop (Maas 1986), was seeded on the area of Field 11 north of the surface drain in the spring of 1995. Yields will be collected in the fall of 1995. Observations show no effects of salinity on crop growth.

Monitoring herbicide leaching. A study was initiated at the drainage site in 1991 to monitor the movement of herbicides through the unsaturated zone (Cessna et al. 1995). Suction lysimeters were installed for soil water collection to monitor herbicide movement in the soil profile. Tile drain effluent samples were collected to estimate herbicide losses below the root zone, an indication of the potential to contaminate ground water. Tile drain effluent residue data showed herbicide movement through the unsaturated zone by preferential paths.

One limitation of the drainage site for herbicide leaching studies is the fact that all tile drain effluent is drained by a common drain conduit, thus, not allowing replication of the data collected. This limitation has been addressed by the installation of a second drainage site at SIDC on a 6.6 ha site with funding from the Canada-Saskatchewan

Agricultural Green Plan Agreement (Tollefson and Hogg 1995). The new drainage site was designed to provide four quadrants that enable data collection of the drain effluent independently. Leaching, preferential flow, and water use efficiency studies will be conducted when both chemical and physical equilibration of this site are complete.

Summary and Conclusion

The reduction in initial salt content and the salt discharge in the effluent, combined with the improvement in crop yield, indicate that successful reclamation of this site has been achieved. Monitoring will continue to ensure that the reduced soil salinity levels are maintained.

The establishment of a specially designed drainage site at SIDC will allow replicated data collection on the movement of herbicides under field conditions. This will provide information for the establishment of best management practices under irrigated conditions allowing for more sustainable agricultural practices.

Acknowledgments

The authors wish to acknowledge Mr. Garth Weiterman, Sask Water, Outlook, Saskatchewan, for his assistance with the EM38-ECa contour map generation using Geosoft, the staff at SIDC for their assistance in soil sampling, EM38 readings, monitoring drain flow rates, and collection of drain effluent samples, and Dr. J. Wahab, Specialty Crops Agronomist, SIDC, for his critical review of this paper.

References

Bennett, D.R., G.R. Webster, B.A. Paterson, and D.B. Harker. 1982. Drainage of an irrigated saline soil in Alberta. *Can. J. Soil Sci.* 62:387-396.

Bole, J.B. 1986. Amelioration of a calcareous solonetzic soil by irrigation, deep ripping and acidification with elemental sulfur. *Can. J. Soil Sci.* 66:347-356.

Cessna, A., L.A. Kerr, T.C. McIntosh, W. Nicholaichuk, J.A. Elliot, K.B. Best, L.C. Tollefson and B. Vestre. 1995. Leaching of herbicides during a fall irrigation using low pressure sprinkler irrigation. In *Proceedings: agricultural impacts on water quality,* February 21-22, 1995. Red Deer, Alberta:179-183.

Chang, C., T.G. Sommerfeldt, J.M. Carefoot, and G.B. Schaalje. 1983. Relationships of electrical conductivity with total dissolved salts and cation concentration of sulfate dominant soil extracts. *Can. J. Soil Sci.* 63:79-86.

Harker, D.B. and D.E. Mikalson. 1990. Leaching of a highly saline-sodic soil in southern Alberta: a laboratory study. *Can. J. Soil Sci.* 70:509-514.

Harron, B. and L.C. Tollefson. 1989. Monitoring salinity after tile drainage. In *Proc. 1989 Soils and Crops Workshop,* University of Saskatchewan, Saskatoon:120-134.

Hoffman, G.J. 1985. Drainage required to manage salinity. *J. Irrig. Drain. Div. ASCE* 111:199-206.

Jensen, N.E. and J.B. Wright. 1986. Investigation of potential drainage project for the Saskatchewan Irrigation Development Centre. Saskatchewan Agricultural Development Fund, Project Number D-86-0010, 14p.

Jones, B.D. and J. Lebedin. 1986. Outlook demonstration farm (SW15-29-08-W3) soil salinity investigation, hydrogeological considerations for drainage. 27p.

Jury, W.A., W.M. Jarrell, and D. Devitt. 1979. Reclamation of saline-sodic soils by leaching. *Soil Sci. Soc. Am. J.* 43:1100-1106.

Maas, E.V. 1986. Salt tolerance of plants. *App. Agric. Res.* 1:12-26.

McNeill, J.D. 1986. Rapid, accurate mapping of soil salinity using electromagnetic ground conductivity meters. Tech. Note TN-18. Geonics Limited, Mississauga, Ontario, Canada.

Rapp, E. 1968. Performance of shallow subsurface drains in glacial till soils. *Trans. ASAE* 11:214-217.

Reeve, R.C. and M. Fireman. 1967. Salt problems in relation to irrigation. In *Irrigation of Agricultural Lands*. Edited by R.Hagan et al. Agronomy Monogram No. 11, Am. Soc. Agron., Madison, WI.

Reeve, R.C., A.F. Pilsbury, and L.V. Wilcox. 1955. Reclamation of a saline and high boron soil in the Coachella Valley California. *Hilgardia* 24:69-91.

Rhoades, J.D. 1982a. Reclamation and management of salt affected soils after drainage. In *Proc. first ann. western provincial conf., soil salinity,* Lethbridge, Alberta. November 29 - December 2, 1982:125-197.

Rhoades, J.D. 1982b. Soluble salts. In *Methods of soil analysis Part 2.* Edited by A.L. Page et al., Agronomy No. 9 Am. Soc. Agron., Madison, WI.

Stushnoff, R. and D.F. Acton. 1987. Soil survey of the Saskatchewan Irrigation Development Centre, Outlook, Saskatchewan. SIP Publication No. M82. 40p.

Tollefson, L.C. and T.J. Hogg. 1995. Irrigation sustainability - Saskatchewan activity. In *Proceedings: agricultural impacts on water quality,* February 21-22, 1995, Red Deer, Alberta:108-118.

U.S. Salinity Laboratory Staff. 1954. Diagnosis and improvement of saline and alkali soils. U.S. Dep. Agric., Handbook No. 60. 160p.

van Schaik, J.C. and R.A. Milne. 1962. Reclamation of saline-sodic soil with shallow tile drainage. *Can. J. Soil Sci.* 42:43-48.

Wither, B. and S. Vipond. 1983. The design of subsurface drainage systems. In *Irrigation: design and practice.* Batsford Academic and Education Limited.

The Reclamation of Prairie Wetlands

by G. Fuller and G. Riemer

Faculty of Engineering, University of Regina, and Saskatchewan Wetland Conservation Corporation, Regina

Abstract. At one time prairie wetlands provided the necessary forage for the mixed farms that were then prevalent on the prairies. Recently the emphasis on grain farming, partially encouraged by some government policies, has reduced the agricultural benefits provided by wetlands. Prairie grain farmers consider wetlands, or sloughs as they often are referred to, as a liability not only because of unrealized seeded acreages, but because circumventing wetlands introduces inefficiencies into farm cultivation operations. Consequently, many prairie wetlands have been drained. These decisions disregard the many benefits wetlands provide to society, such as groundwater recharge, flood control, erosion control, and wildlife habitat. As drainage continues, the remaining wetlands become more valuable to society. Since society benefits from wetlands, society must accept responsibility for preserving wetlands. Legislation has been passed to preserve wetlands, but the drainage of wetlands continues. As a result, future generations may not enjoy some of the benefits wetlands provide. This paper discusses the value of wetlands as well as the reasons that wetlands continue to be drained, and gives suggestions for the reclamation of wetlands.

Introduction

Farmers are maximizing crop production to offset their increasing costs. Maximization of crop production has included the drainage and cultivation of wetlands and their peripheral natural vegetation. Present day farmers consider natural wetlands to be a liability not only because of unrealized crop production but also because circumventing the wetlands introduces inefficiencies into the farming operations. Farmers also benefit from the cultivation of wetlands because grain quotas generally are based on cultivated acreages. Hence, marginal lands have been cultivated primarily for the additional quotas they provide (Zittlau 1979).

Benefits of Wetlands to Society

Although some individual farmers may profit from draining wetlands, society in general enjoys many benefits from natural wetlands. The drainage of wetlands increases runoff peaks and the discharge of farm chemicals downstream. This situation contributes to downstream flood damage and to eutrophication. The drainage of wetlands destroys wildlife habitat which has human impacts too, affecting not only naturalists but trappers and hunters as well. Wetland drainage also reduces the amount of water available for recharging aquifers, which may result in future water supply shortages. The loss of one small pond does not have a significant effect on the hydrology of a region and, since decisions often are made on the basis of individual wetlands, these are drained without much assessment of the impacts of their drainage. However, the cumulative effects of the loss of many wetlands throughout a region may cause large economic losses as a result of flooding, erosion, and water supply shortages.

Society enjoys many recreational activities that depend upon wetlands. These activities include hunting, fishing, hiking, bird watching, canoeing, camping, and photography. Due to the recreational opportunities associated with wetland areas, most parks and picnic sites are located adjacent to wetlands and their peripheral habitats. Large sums of money are spent on recreational activities associated with wetlands. Each prairie wetland that is drained reduces our capability for maintaining waterfowl populations. Approximately one half of all North American waterfowl are raised on the Canadian prairies (Lynch-Stewart 1983). As wetlands are drained and waterfowl are restricted to fewer areas, the possibilities of losses due to disease and predators increase and the numbers of waterfowl decrease. Society is becoming more concerned about the continued draining of the remaining wetlands. Nevertheless, the practically irreversible decisions to destroy wetlands and their surrounding vegetation continue. Millar (1981) has estimated that by 1979, 84% of the wetlands in southern Saskatchewan had been affected by human activities.

The benefits that wetlands contribute to society often are not considered when decisions regarding drainage are made because these decisions often are based primarily on the production capability of individual farms. Therefore, it is not surprising that the drainage of wetlands continues because farmers generally own the lands that contain the wetlands, and their objective is to produce agricultural products. Increasing the cultivated acreage of a farm for a reasonable drainage cost may be a sound farm business decision based on the costs and benefits to the farm. However, the resulting costs to the region, such as flood control, may be greater than the individual benefits derived from increased crop production.

The Land Use Conflict

The land use conflict between agricultural drainage and wetlands conservation involves physical, socioeconomic, and political interrelationships. Some benefits of wetlands are not included in an economic analysis because the benefits cannot be expressed in monetary terms. In the past farmers have borne the costs of wetlands but there has been no mechanism for them to market the contribution their wetlands provide to society. On the other hand, society has enjoyed benefits from wetlands but there has not been a suitable market mechanism for them to express their preference for the benefits of wetlands, such as wildlife. Political problems with respect to drainage result from the fact that the surface water and wildlife associated with wetlands generally are located on private lands. Governments have the conflicting responsibilities of improving the social welfare of society and protecting the individual rights of farmers. To further complicate the matter, governments may be providing assistance for the drainage of wetlands while at the same time providing assistance to improve wildlife habitat.

Since it is society in general and not just the farmers that enjoy the benefits of wetlands, it follows that society should accept the financial responsibility of preserving these resources by reimbursing the farmers for not destroying the wetlands and their peripheral vegetation. This could be accomplished by a government rebate of municipal land taxes for retaining wetlands. A drainage tax or increase in land tax assessment also could be levied on wetlands that are drained to discourage the drainage and cultivation of marginal lands, since the drainage of such lands generally is not beneficial to either farmers or to society. To discourage the drainage and cultivation of marginal lands, cultivated acreages should not be the criterion for allocating quotas. Other factors such as land fertility, as indicated by the land assessment, should be considered.

Society is concerned about the rate at which wetlands and their peripheral vegetation are being destroyed. Nevertheless, the destruction continues, sometimes aided by our governments. The present approach of making drainage decisions on the basis of individual wetlands and farms must be replaced by a more regional approach that considers all of the regional effects of drainage. This will give a better representation of the value society places on wetlands and a better assessment of drainage impacts. Otherwise, the benefits that wetlands provide to our society may not be enjoyed by future generations.

References

Lynch-Stewart, P. 1983. *Land use change on wet lands in southern Canada: review and bibliography.* Environment Canada, Lands Directorate, Ottawa.

Millar, J.B. 1981. *Habitat changes in Saskatchewan waterfowl strata 30 to 33 between fall 1978 and fall 1980.* Prairie Waterfowl Habitat Evaluation Program, Canadian Wildlife Service, Saskatoon, Saskatchewan.

Zittlau, W.A. 1979. *An environmental assessment of agricultural practises and policies: implications for waterfowl management.* Natural Resources Institute, University of Manitoba, Winnipeg.

Reclamation for Multiple Land Use at Big Quill and Chaplin Lakes: Case Studies in Building Partnerships

by Greg Riemer

Manager, Agricultural Services, Saskatchewan Wetland Conservation Corporation, Regina

Abstract. As part of the Saskatchewan commitment to the North American Waterfowl Management Plan, the Saskatchewan Wetland Conservation Corporation (SWCC) is acquiring over 30,000 acres of native vegetation, some of it in poor condition, in a contiguous block around Big Quill Lake in east central Saskatchewan. This lake is of international significance to waterfowl and to nesting and migrating shorebirds. SWCC goals for this area are to improve the condition and vigour of the range for livestock and to provide and protect habitat for wildlife such as Baird's Sparrows and Piping Plovers.

Introduction

Big Quill Lake is a shallow, saline lake whose waters and mud flats cover approximately 90 square miles. It is a major migratory stopover for shorebirds, having been dedicated recently as a Western Hemisphere Shorebird Reserve Network (WHSRN) site. Prior to Saskatchewan Wetland Conservation Corporation (SWCC) involvement, approximately half of the 30,000 acres was accessible to cattle for season-long grazing, whereas the other half was vacant but had received some grazing in the past from unauthorized free-ranging cattle. A history of land use was obtained, a range evaluation was carried out, and management plans were developed and implemented. An extensive inventory of the range and its condition was done to develop multiple use management plans. These plans recommended cattle access be removed from approximately 20 miles of shoreline and 7000 acres with vulnerable forage resources.

Water development and fence construction on 8400 acres on four separate pastures will allow the implementation of deferred and rest-rotation grazing systems. Grazing will be deferred on another 870 acres of riparian habitat. Improved cattle distribution and better utilization of forage resources should allow local livestock producers to maintain a similar number of cattle as in the past and to improve vegetative cover for grassland-nesting birds. The catalyst for this work and the many partnerships that were developed through it was the United States' North American

Wetland Conservation Act (NAWCA), which funds 75% of the North American Waterfowl Management Plan. The initial contributors include the states of Tennessee, Kansas, Nebraska, and Wyoming, as well as the Nature Conservancy and Wildlife Habitat Canada. Local partners include the Canadian Wildlife Service (CWS), Prairie Farm Rehabilitation Administration (PFRA), Ducks Unlimited, Saskatchewan Power Corporation, local landowners, and grazing cooperatives.

SWCC has begun preliminary work on the Chaplin Lake complex. Like Big Quill Lake, shorelines provide prime nesting habitat for the Piping Plover, an endangered species. Both places are also within the range of Baird's Sparrow, a threatened species known to prefer lightly or ungrazed native grasslands. During the peak of the migration in late May, Chaplin Lake is a stopover for one-half of the entire population of Sanderlings.

Chaplin Lake is mined for sodium sulphate and its brine shrimp. Its lunar appearance is misleading, as Chaplin Lake is rich in tourism opportunities for bird watching and other natural area interpretation. An ecotourism study was initiated with the assistance of Gold Corp and the PFRA's Partnership on Agriculture and Rural Development (PARD). Range evaluation and grazing system designs have been done with the financial assistance of the NAWMP and CWS. Shorebird studies have begun to quantify numbers and species for the lake's WHSRN dedication.

This paper reviews the rationale, evaluation, planning, and partnership building required to implement both projects.

Background

The corporate mission of the SWCC is to lead and coordinate the province's wetland conservation initiatives to ensure the sustainability and biodiversity of the prairie environment for people and wildlife.

SWCC was created on January 10, 1990 to, among other things, coordinate provincial activities on behalf of Saskatchewan North American Waterfowl Management Plan (NAWMP) partners. The NAWMP, a continental partnership among Canada, the United States, and Mexico, seeks to restore waterfowl and other wetland dependent wildlife populations through sound land use programs that contribute to soil, water, and wildlife conservation. (For more detail on the NAWMP see the Appendix to this paper.) Within this context, SWCC is making a significant contribution to achieving objectives of Saskatchewan's Environmental Strategy by maintaining and improving biological diversity, soil conservation, and water quality and quantity.

By protecting the province's wetlands and associated uplands and by encouraging wildlife-friendly agricultural practices, the corporation is making impressive strides toward achieving its NAWMP objectives. This is being accomplished through innovative partnerships at local, regional, and international levels.

SWCC is guided by a Board of Directors with representation from provincial and national organizations that affect land use in the province. Board membership spans both private and public organizations in the agricultural, wildlife, and environmental sectors. This mix brings diversity and broad insight to the activities of the corporation. SWCC represents a partnership of provincial and national wildlife habitat conservation and land use agencies. Through this partnership, the corporation uses an integrated land use approach by linking agricultural and wildlife interests to NAWMP programming in Saskatchewan.

In the past years the NAWMP and the corporation have used an ecosystem approach to improve wildlife habitat in general, and waterfowl, shorebird and grassland songbird habitat in particular. Clearly, what is good for the ecosystem is good for waterfowl. The net result will be an umbrella of protection for wildlife that rely on wetlands, uplands, and the enhancement of a resource which greatly benefits our province's agricultural and wildlife-related industries.

Management Objectives

The land use goals of the Saskatchewan Wetland Conservation Corporation are to:

- preserve habitat for nesting and migrating shorebirds;
- preserve habitat for grassland birds, such as Baird's Sparrow and the Western Meadow Lark;
- improve range condition and pasture production for local cattle producers; and
- enhance the overall resource to improve its potential for tourism to stimulate local economies.

Multiple Use Management at Big Quill Lake

Rationale for land acquisition and improvement. The land in the Quill lakes region was surveyed initially in the 1890s. At that time Big Quill Lake was much larger and the surveys stopped at the existing shoreline. In subsequent decades the lake level dropped, never to fully recover. This created over 30,000 acres of unsurveyed land, in essence a "no man's land." Approximately 40 years ago permanent vegetation had become established and grazing had started under the auspices of the provincial land agency. The present grassland resource is based on a saline soil substrate, and years of poor grazing management have resulted in poor to fair range conditions, with large variations in animal use. The rangeland around the lake also was in poor condition for wildlife habitat and required reclamation. Unrestricted cattle access to the shoreline was a major concern if the area was to be effectively managed for the endangered Piping Plover.

Prior to the signing of the NAWMP, a pilot project was set up by the major partners of the still unnamed NAWMP to look at broadly based landscape management to aid waterfowl nesting. Waterfowl biologists had determined earlier that the critical element limiting waterfowl populations was loss of upland nesting cover, which throughout most of the prairies was being converted to annual crop land. The place chosen for the original pilot project was the Quill lakes area. Its landscape contains a large saline lake, a large freshwater lake and thousands of small wetlands. While adversely impacted by cultivation and overgrazing, Big Quill Lake was a very important staging area and migration stop for approximately half a million geese and cranes (with one-day counts in October 1993 in excess of 300,000 geese), 150,000 arctic shorebirds, and a small number of Whooping Cranes. The saline shoreline of Big Quill Lake is the nesting habitat of about 300, or 5%, of the world's remaining population of Piping Plovers. The grassland complexes around the Quill lakes are important to many ground-nesting shorebirds, including Baird's Sparrow, which is considered threatened.

In the 1980s, the RAMSAR Convention designated the Quill lakes as a wetland complex of world significance. Recently, in cooperation with the Canadian Wildlife Service of Environment Canada and Saskatchewan Environment and Resource Management, SWCC nominated the Quill lakes as a Western Hemisphere Shorebird Reserve Network site. These designations have provided international recognition of the value of this site for migrating birds, as well as increased opportunities for development of the province's ecotourism industry.

Range assessment and management plans. A range assessment (Abouguendia 1990; Wroe et al. 1988) was performed on approximately 32,000 acres adjacent to Big Quill Lake. The area is located in the black soil zone and for the most part has been classified as a saline lowland range site. Range condition varies from poor to low good, with the dominant plant species of the area being northern reed grass *(Calamagrostis inexpansa)*, slender wheatgrass *(Agropyron trachycaulum)*, saltgrass *(Distichlis stricta)*, foxtail barley *(Hordeum jubatum)*, Nuttall's alkali grass *(Puccinella nuttalliana)*, bluegrasses *(Poa* spp.*)*, many-flowered aster *(Aster* spp.*)*, sowthistle *(Sonchus* spp.*)*, gumweed *(Grindelia squarrosa)*, goldenrods *(Solidago* spp.*)* and red samphire *(Salicornia rubra)*.

In many areas, cattle have had access to shorelines, beaches, and mud flats with poor vegetative cover, and uneven cattle distribution has resulted in increased grazing pressure along a major creek flowing into Big Quill Lake, near water sources, and along higher ridges. The saline environment also appears to have been responsible for patchy vegetative cover in several locations.

Management plans (see pasture plans for Wimmer Grazing Co-op, Lampard Grazing Co-op, Berlinic and Pruden Pastures) recommended stocking rates ranging from 0.24 to 0.44 Animal Unit Months per acre and employed fencing and water development to: (i) eliminate grazing on 16 miles of shoreline and 12,000 acres of mud flats and beaches, (ii) eliminate grazing on 7000 acres of vacant lands with patchy vegetation, (iii) allow the implementation of rest-rotational and deferred grazing systems on 8400 acres, and (iv) defer grazing on 870 acres of riparian habitat. Two reports, "Range Evaluation and Grazing Management Plans for Leased Pastures in the Big Quill Area" and "Range Evaluation and Grazing Feasibility for Vacant Lands in the Big Quill Lake Area" were written by Sweet Grass Range Consulting. Both reports currently are available from the SWCC.

Partnerships. Changes made to the Wildlife Habitat Protection Act and new regulations added to both it and to the provincial Lands Act in 1993-94 permit the transfer of Crown lands from Saskatchewan Agriculture and Food to the SWCC. These wetlands and adjacent uplands will help meet provincial commitments to the NAWMP. Under the terms of Saskatchewan's agreement to join the NAWMP, Saskatchewan Agriculture and Food is providing half of the province's commitment to the plan as Crown land.

The implementation of the grazing plans requires the consent of the pasture lessees as they surrender lease with Saskatchewan Agriculture and Food and sign a new lease with SWCC. To acquire this land SWCC must have it surveyed. Arrangements have been made with Saskatchewan Justice to survey the parcel in large continuous blocks and not in the familiar British mile square sections and 160-acre quarter sections. This new arrangement will create a contiguous block of land with restricted access as there are no road allowances. Transfer on the first holding, a 3758-acre parcel at the north end of Big Quill Lake was completed in the winter of 1994. This transfer was the first of almost 20,000 acres of vacant land and 12,400 acres of leased pastures.

To construct the pasture improvements, SWCC utilized funding from the Canadian Wildlife Service of Environment Canada under its work plan for the NAWMP. Additional funding to improve the Dafoe pasture and evaluate and acquire large tracts of vacant land was obtained from the PRAIRIE SHORES program which is delivered by SWCC through NAWCA funding. PFRA was looking for a demonstration of managed grazing of riparian systems and funded the riparian development of the Wimmer Grazing Co-op. Ducks Unlimited Canada had some of its program delivery money earmarked for endangered species work and generously consented to fund the construction of seven miles of shoreline fencing to exclude cattle from the Piping Plover nesting areas.

The improvement of pastures required fencing and rotational grazing systems to better utilize the grass resource. This required extensive improvements to the cattle watering systems. Water had to be piped from existing scarce dugouts as far as two miles. Saskatchewan Power Corporation entered into an agreement with SWCC to provide solar-powered pumping units for seven of the sites that did not have access to the power grid. The benefits are numerous as power poles and overhead lines are not required in a major staging area, the rangelands have not been impacted by heavy equipment, and an alternative power source is demonstrated. These systems now are up and running and in several cases providing ample water for well over 200 head of cattle. Sask Power's involvement was motivated by their vision for environmentally friendly electrification of a world class resource. For its substantial contribution Sask Power received permanent recognition on large informational highway signage, signage at habitat access points and where the solar units were located, and in tourism brochures for the Quill lakes.

Two land acquisitions helped consolidate wildlife habitat around Big Quill Lake. The Saskatchewan Wildlife Federation made a gift of 480 acres to SWCC, and the corporation purchased one quarter section of key habitat along the south shore with the financial assistance obtained from its Habitat Diversity Program.

Chaplin Lakes Habitat Enhancement and Ecotourism Development

Goal. The long-term goal at Chaplin Lake is to provide and protect shoreline habitat for migratory shorebirds. This development process has created opportunities for a major ecotourism business in Saskatchewan as well as the village of Chaplin. Of key importance here is the securement of shoreline areas for nesting Piping Plovers and migrating shorebirds, and improvement of surrounding rangelands for ground-nesting birds. This project will sustain and enhance the rangeland-based livestock industry, provide local jobs, help diversify the local economy, and provide residents the opportunity to establish a new sustainable tourism product.

Rationale for land acquisition and improvement. Chaplin Lake is a large saline lake between Moose Jaw and Swift Current on the TransCanada Highway. Prior to its development it was dry most of the time and of little value to wildlife or to the people of the area. Its reclamation for agriculture would have been prohibitively expensive and the results of such reclamation work have had low success. What we now have at Chaplin Lake is the utilization of a saline ecosystem to its highest and best use. With the involvement and financial support of Ducks Unlimited and Saskatchewan Water Corporation, water has been diverted to the lake to create a large freshwater marsh at the southeast end of the lake and provide water to the north end of the lake which is easily visible from the TransCanada Highway. This water is utilized by Gold Corp to "brine" the lake to enable the mining of sodium sulphate. The mining process has had some very unexpected and beneficial results. The large shallow cells in the lake are the very best possible way to encourage the growth of brine shrimp and other invertebrates. Like Big Quill Lake, these saline conditions are perfect feeding sites for arctic-nesting shorebirds. The mining operations have created an almost consistently perfect migratory stopover for shorebirds. The lake probably has been used by shorebirds for thousands of years, but only in those infrequent years when there was enough runoff to fill it.

Shorebird surveys conducted by the SWCC resulted in a count of 67,000 birds of varying species using the area in a single day. Fifty thousand Sanderlings were counted, which represents up to 50% of their hemispheric population. The endangered Piping Plovers also use the shoreline for nesting and raising their young.

Partnerships. Through the efforts of SWCC and the Canadian Wildlife Service, Chaplin Lake will be designated a "Hemispheric" Western Hemisphere Shorebird Reserve Network site in May of 1996. This is the highest designation possible.

SWCC and Nature Saskatchewan, using seed money from Gold Corp, received approval November 23, 1994 to develop a marketing/development strategy for Chaplin Lake under the Canada-Saskatchewan Partnership Agreement on Rural Development (PARD). This strategy followed SWCC's involvement in the Provincial Ecotourism Study, which was cost-shared with Saskatchewan Economic Development.

Red Pel Corporation, a Saskatchewan ecotourism consulting firm, was contracted to undertake the study and have it completed by July of 1995. The study listed a number of developments that are required for the project, identified key nature-viewing sites, outlined a marketing and promotion plan, and made recommendations as to how local residents could develop the area. SWCC continues to expand partnerships to include local and provincial tourism development and marketing interests.

SWCC, utilizing CWS and PRAIRIE SHORES funding, has undertaken range evaluation and planning of the pastures surrounding Chaplin Lake. The corporations' objectives are the same as at Big Quill Lake, namely, to design and implement multiple use grazing systems that enhance the rangelands around the lake for both livestock and wildlife.

Working with Gold Corp and Sask Water, SWCC hopes to ensure that sufficient water is available to ensure further that adequate feedstocks are there for the huge numbers of shorebirds that arrive in the spring.

Expected Benefits

Once management plans have been fully implemented and the systems are operational, several benefits are anticipated:

1. There should be an improvement in the range condition and vegetative cover resulting from improved grazing management.
2. There will be a reduction in conflict between cattle and nesting shorebirds, especially on the beaches where the Piping Plovers nest.
3. Improved range condition and better cover should improve nesting habitat for grassland nesting birds.
4. There will be large areas adjacent to pastures that will be set aside for wildlife only. In addition, a portion of the paddocks within the pastures will not be grazed until the nesting season is over and at least one paddock per pasture will not be grazed each year.
5. There will be a sustainable supply of forage for cattle for local livestock producers.
6. Opportunities will be created for development of an ecotourism service industry.

Acknowledgments

Sponsorship of the project is provided by: PFRA, which funded the riparian fencing on the Wimmer Pasture; Ducks Unlimited, which funded the construction of four miles of cattle-exclusion fencing along the shorelines; the Saskatchewan Power Corporation, which constructed the solar water-pumping systems and drafted the plans used in this paper; the Canadian Wildlife Service, through work plan funding of SWCC's land programs; the NAWCA, through its funding of the PRAIRIE SHORES Project; PARD and Gold Corp, for their funding of the Chaplin ecotourism development and marketing study; and the local communities for support, vision, and the work to make it happen. The authors wish to thank Mr. Garfield MacGillivray, Wilderness Images, P.O. Box 386, Quill Lake, Saskatchewan and Dave Gleim, P.O. Box 61, Chaplin, Saskatchewan, for providing many of the wildlife and pasture slides used in the presentation.

References

Abouguendia, Z.M. 1990. *Range plan development.* Saskatchewan Research Council.

Harrison, Tom. 1993. *Range evaluation and grazing management plans for leased pastures in the Big Quill area.* Saskatchewan Wetland Conservation Corporation, Regina.

Harrison, Tom. 1993. *Range evaluation and grazing feasibility for vacant lands in the Big Quill Lake area.* Saskatchewan Wetland Conservation Corporation, Regina.

Wroe, R.A., S. Smoliak, B.W. Adams, W.D. Willms, and M.L. Anderson. 1988. *Guide to range condition and stocking rates for Alberta grasslands.* Edmonton, Alberta.

Appendix

North American Waterfowl Management Plan Structure and Programs

The North American Waterfowl Management Plan (NAWMP) is an agreement among Canada, the United States, and Mexico to cooperate in restoring waterfowl populations to the levels of the 1970s and to improve habitat for other wetland-dependent wildlife.

The NAWMP encourages and helps focus investment by wildlife conservation organizations from across the continent to critical habitat areas for migratory birds. Through the NAWMP, Saskatchewan, Alberta, and Manitoba comprise the Prairie Habitat Joint Venture (PHJV) for protecting key waterfowl habitat, one of 12 habitat joint ventures in North America. The PHJV is the NAWMP's top priority because it provides breeding habitat for almost 40% of the continent's duck population, including 50% of Mallards and over 55% of Pintails.

In Saskatchewan, NAWMP activities are coordinated by the Saskatchewan Wetland Conservation Corporation (SWCC). SWCC also represents provincial interests on the PHJV advisory board. Provincial NAWMP partners guide the corporation's activities through their representation on SWCC's board of directors. Programs contributing to provincial NAWMP objectives include Large Marsh, Nest Baskets, Prairie CARE, PRAIRIE SHORES, and Waterfowl Crop Damage Prevention and Compensation. Delivery agencies and mechanisms vary with each program.

Minimizing Impacts of Highway Construction on Freshwater Wetlands in Nova Scotia

by Norval Collins and Lynn Davis

CEF Consultants Ltd., Halifax, Nova Scotia

Abstract. In the past, highway designers and builders in Nova Scotia showed little concern about potential drainage to freshwater wetlands on their routes. They focussed on ensuring that water was directed away from the roads, rather than on protecting or establishing wetlands; in the process, they destroyed many wetlands, while inadvertently creating others. Recently, highway builders have become more aware of the value of wetlands, and have been taking a fresh look at the way they deal with them, as have many other project proponents. Two major questions need to be addressed to determine the overall potential for impact of a highway on wetlands. First, what are the characteristics of the wetlands within the potential impact area? Second, what is the "value" of these wetlands within the potential impact area? The first question can be addressed by specific field studies within the impact area. The second question is more difficult to address if available information is limited, as is the case for many Nova Scotia wetlands. Satellite image analysis provides one means of identifying other wetlands with similar characteristics in the area. Highway construction impacts may be avoided by realigning the proposed route, usually the most extreme solution, or by specific design and construction measures that mitigate potential impacts. In cases where it is impossible to avoid damaging smaller wetlands along the highway right-of-way, lost habitat may be replaced with created or restored wetlands at relatively low cost. It is, however, extremely difficult to duplicate the characteristics and functions of a natural wetlands. Therefore, wetland creation should be considered an additional compensation measure, rather than a substitute for ensuring that existing wetlands are not harmed as a result of construction projects.

Introduction

Highway designers and builders in Nova Scotia have not always shown much concern about damage to freshwater wetlands on their routes. They have tended to focus on ensuring that water was directed away from the roads, rather than on protecting or establishing wetlands. In the process, they destroyed many wetlands, while inadvertently creating others. Recently highway builders and others have become aware of the value of wetlands and have been taking a fresh look at the way they deal with them. This paper addresses this process.

Characterization and Valuation of Wetlands

Two major questions need to be addressed to determine the overall potential for impact of a highway on wetlands. First, what are the characteristics of the wetlands within the potential impact area? Second, what is the "value" of these wetlands in relation to those in the surrounding areas? The first question can be addressed by specific field studies within the impact areas, including typing of wetlands, definition of plant and animal assemblages, water quality sampling, and hydrologic monitoring and analysis. The second question is more difficult to address if available information is limited, as is the case with many Nova Scotia wetlands.

Characterization. Wetlands may be characterized by many features, but dominant plant species and water quality are those most commonly used. For example, bogs frequently are classified by the amount of tree cover. In a recent study, we initially classified bogs into treed and untreed types; however, further investigations indicated that all of the bogs were treed to some extent, making this division impractical. After analysis, bogs were typed by dividing them into:

- sphagnum bogs, where approximately 85% of the wetland was covered by sphagnum;
- sedge bogs, where 60% of the surface was covered by sphagnum with sedges covering about 30% of the surface; and
- treed bogs, where about 10% of the wetland was treed and lichens predominated in the ground cover.

All bogs in the study area are dominated by sphagnum moss. Further successional stages are reflected by the invasion of trees such as black spruce and larch, and ericaceous species along the edges. A treed bog may still have an open centre. Advanced stages may be completely forested or may be relatively dry ericaceous barrens (raised bogs) with abundant lichen cover on the ground. Some trees or large shrubs are present in all bogs investigated within the proposed highway corridor, but the classification of a treed bog has been reserved for the drier sites with abundant lichen cover. In addition, two wetlands initially classified as bogs have been reclassified as sedge meadows following field investigations because of the relatively low (20%) surface coverage by sphagnum.

Water quality and hydrology also are used to classify wetlands. Particularly important parameters include dissolved nutrients, salt concentration, pH, dissolved oxygen, conductivity or total dissolved solids, total organic carbon, and colour (Hammer 1992).

Key factors in analysis include the degree of acidity affecting the plants that are dominant, and indicators of the organic acids present, provided by total organic carbon, pH, and conductivity. Since wetlands often act as a "sink" for many chemicals because complex organic acids tend to bind ions, an indication of the quantity and types of organic acids present can be useful in classification. An understanding of the hydrology or flows within the wetland is useful, but frequently difficult to obtain because of its complexity.

There is no single correct classification for wetlands in a study of this type; the system adopted should be the one most appropriate for addressing the issues involved.

However, it is essential that a range of data be collected to ensure that an appropriate system can be developed.

Using 1:10,000 orthophoto mapping with updating from air photos, wetlands may be initially grouped into bogs, fens, and marshes. Field surveys then are carried out to determine the important characteristics of each of these basic types within the study area. Information generally is collected by detailed identification of all vascular plants within random quadrats, with percentage cover by families of lower plants noted. Transects near the proposed centreline also are used to assess the potential for uncommon or rare endangered plants within the proposed right-of-way. The timing of surveys is critical, with many species only identifiable during their flowering stage.

A report on rare or endangered species of plants or animals in Nova Scotia provides a listing of known locations of these species in the province (Isnor 1981). As well, up-to-date listings of rare and endangered species from within the region can be obtained from the Nova Scotia Museum and are essential to have during field studies. They can provide an aid in identification, allowing a botanist to recognize a plant as a possible rare species even if it is in a vegetative state and can only be confirmed to species when in flower.

Data collection requirements for fish, birds, and other animals vary depending on the area. For example, regional information on breeding birds in Nova Scotia is available from a recently completed breeding bird atlas (Erskine 1992). This atlas provides a list of birds observed within 10 kilometre squares according to the standard UTM grid. These data provide a means to evaluate the potential for uncommon bird species to be found within the potential impact area and can be used to determine what further study may be required. The *Nova Scotia Wetland Atlas* also provides an indication of the importance of an area to waterfowl. All wetlands of measurable size are ranked in the atlas with numerical scores. Wetlands scoring greater than 65 are considered to have significant waterfowl habitat value, with scores of 65 to 70 representing good habitat, 70 to 80 representing better habitat, and greater than 80 representing the best habitat.

Similarly, baseline information on at least commercially important species of fish and other animals (e.g. furbearers and big game) are available from government agencies. Field studies generally include electrofishing to document any fish present, and reconnaissance of the area looking for signs of wildlife, including characteristic feces, burrows, signs of browsing, beaver dams, etc.

Once this information has been assembled and reviewed, specific questions may be raised leading to further, more detailed studies. For example, in the study mentioned previously, we focussed further field work on amphibians because some uncommon species were thought to be present in the area, but no data were available to confirm this.

Valuation. Once sufficient information is available to characterize the wetlands within the highway corridor, it is possible to attempt to determine their value in relation to other wetlands in the region.

Satellite image analysis provides one means of identifying other wetlands with similar characteristics in the area which can then be used as a basis for comparison. Satellite imagery is relatively expensive, generally about $2500 for a scene, but it provides a means to evaluate regional characteristics based on recent information. If wetlands within the study corridor have been characterized through initial studies, image analysis then can provide a means for identifying other local areas with similar characteristics. This method has been used in our study to select sites for further evaluation in the region.

A spot satellite image from June 1992, with a resolution of 20 metres, was used to identify wetlands outside of the corridor that had similar spectral characteristics to those within the corridor. The date of this image conformed to the period of the year when field studies focussed on amphibians were to be conducted. First, the spectral characteristics within the corridor were examined. These wetlands had been identified from 1:10,000 orthophoto mapping, examination of 1985 colour and 1991 black-and-white aerial photography, and field visits. Spectral analysis then was carried out to identify other areas with similar characteristics outside of the corridor. The availability of access points for each site then was used to select the final sites for field reconnaissance. Every effort was made to select sites similar in size and appearance to those within the corridor.

Satellite imagery also can be used to provide an illustration of the regional biophysical features of the area. The image can be smoothed to produce a sharp image that presents the most important information clearly, e.g. the broad regional occurrence of wetlands. This smoothing improves the clarity of the image but reduces its spatial resolution. This means that small features of less than a few hundred metres diameter are lost. In particular, many of the small wetlands which have been identified in initial images used for selection of study sites were not visible in the final smoothed image.

Further evaluation of sites both within and outside the highway corridor provided a basis for judging the significance of wetlands characteristics in a regional context. For example, although an uncommon species of salamander was found at sites within the highway corridor, the same species also was identified in wetlands outside the corridor.

Mitigation

The most drastic mitigatory measure is that of avoidance, i.e. realignment of the proposed route. A realignment would be recommended if the impacts were so severe that no mitigatory measures could alleviate the damage, and no positive effects could compensate for the loss. The presence of rare or endangered species of flora or fauna, the destruction of whose habitat would cause a serious decline or loss of the species in the area, would provide sufficient reason to recommend moving the highway .

However, any shift in alignment would affect a length of highway proportional to the distance moved from the centreline because of requirements to minimize the radius of impacts in other areas along the right-of-way.

Design and construction. Highway designers need to take into account potential impacts on wetlands from both construction and operational phases of the highway.

Unlike many streams, wetland areas are not usually characterized by extreme slopes which accelerate runoff. Nonetheless, sedimentation of streams and wetlands is one of the most potentially serious impacts of highway construction. Care must be taken during the construction phase to ensure that silt-laden runoff does not contaminate either the wetlands near the right-of-way or the streams feeding into them. Erosion and sediment control measures must be thoroughly planned in advance of construction.

Construction activities should be restricted to low risk periods of the year. For example, clearing should be carried out in winter after the ground is frozen to minimize damage by heavy equipment. Clearing in winter also minimizes the impacts of breeding activities of birds and other wildlife.

Careful design and construction also are critical for stream crossings to prevent sedimentation and disruption of water flow to wetlands. Stream crossings should be constructed in summer under low flow conditions, and should be completed prior to mid-October when salmon spawning takes place in Nova Scotia streams. Construction equipment should not be permitted in water courses. Fisheries and Oceans Canada reviews erosion control measures and construction designs for all stream crossings during environmental review of the detailed construction program.

Culverts often are placed under highways which cross small fens. The vegetation and hydrology of fens, however, can be dependent on subsurface water flows. When this is the case, crossings using culverts can result in habitat changes for an unpredictable distance from the crossing. Impacts on fens usually can be mitigated by constructing a bridge rather than via a culvert crossing, but the length of span required will depend on the hydrology of the particular fen and the distance required to avoid restricting subsurface flows. The balance needs to be addressed during detailed design of the crossings.

Bogs are extremely sensitive to hydrologic changes. Thus, special care should be taken to ensure that construction activities in bogs cause minimal disruption to the wetland. Clearing should take place during winter on frozen ground and construction should occur in summer when water levels are low. The construction site should be isolated from the bog, with work times arranged so that water drawdown occurs for the shortest possible time to avoid stress on wetland flora.

Because of the importance of some wetlands to the hydrology of the surrounding area, it may be critical that the existing hydrologic regime be preserved as much as possible. One option is to place a series of smaller culverts at intervals where the highway intersects with the bog, rather than using large culverts or filling in the bog entirely. For example, the Minnesota Department of Transportation proposed installation of 24-inch culverts spaced at 100-foot intervals as a way of preserving a moss wetland (Adamus and Stockwell 1983). More numerous, smaller culverts also increase the chance that the flow capacities can be modified to resemble natural conditions.

Once the highway is operational, it is important that point-source drainage from the highway be avoided to reduce hydrologic impacts on the surrounding wetlands. Rather than concentrating runoff into a single channel along the highway, drainage should be provided away from the highway in a number of locations, and flow dispersed into undisturbed natural vegetation at minimal velocities.

Despite the most careful construction, the natural water regime of wetlands within the highway corridor will be inevitably disrupted to some extent as a result of construction activities and highway operation. As a result, a certain amount of wetland management may be needed, especially in the first few years after construction. Wetland management primarily consists of monitoring and manipulating water levels. This usually is accomplished by using water control structures, which may be as simple as a stop-log system (Hammer 1992) or sand bags. Modification to water control structures to minimize blockage by beavers also has been proposed (Hammer 1992). Most authorities refer to the importance of monitoring, accompanied by the ability to make adjustments to water flows and levels, as being critical to the long-term success of created or altered wetlands (Kusler and Kentula 1990; Lowry 1990; Hammer 1992).

Creation of Wetland Habitat

In the United States, section 404 of the Clean Water Act requires restoration, enhancement or creation of wetlands to offset unavoidable construction impacts that cannot otherwise be minimized (Kusler and Kentula 1990). Most authorities agree that restoration should be favoured over creation, chiefly because the landscaping and design requirements usually are less complex (Kusler and Kentula 1990; Garbisch 1986).

Much of the wetland creation which has taken place has involved coastal wetlands, waterfowl marshes, and impoundments. Ducks Unlimited has been engaged in wetland creation projects for years. Its chief aim, however, is to create waterfowl habitat, which is not necessarily of the most benefit to other species, such as amphibians. Most Ducks Unlimited projects are directed toward creating hemimarshes, which consist of half open water and half vegetation dispersed throughout the water. Marshes for waterfowl habitat require a sufficient supply of nutrients to promote plants and invertebrates to be used as food. In addition, a suitable water depth, usually less than one metre, is needed to make the area attractive to ducks (Ducks Unlimited Canada, n.d.).

Many of the wetlands in Nova Scotia are quite different from the type Ducks Unlimited tries to create. For example, treed bogs or sphagnum bogs have very little open water. They characteristically have low nutrient levels, and thus do not produce the sort of vegetation suitable for waterfowl habitat. Much less work has been done on the establishment of bogs, fens, and forested wetlands which have extremely sensitive long-term hydrologic requirements (Kusler and Kentula 1990; Zedler and Weller 1990; Garbisch 1986). However, although experts agree that *total* duplication of a natural wetland is impossible, some systems may be approximated and certain wetland functions may be created or enhanced (Kusler and Kentula 1990). Hammer (1992) suggests that since marshes are relatively easier to establish and are successional stages to bogs, efforts to create bogs could begin by establishing marshes.

Design criteria. Freshwater wetlands are extremely sensitive to hydrological conditions, especially in the early stages of development (National Wetlands Working Group 1988). Therefore, an understanding of wetland hydrology and the ability to adjust and manipulate the water regime is critical to the success of a wetland creation or restoration project (Kusler and Kentula 1990; Lowry 1990; Hammer 1992).

Borrow pits and spoil areas along highway rights-of-way offer opportunities for wetland creation at relatively low cost. Rather than grading borrow areas to a uniform elevation, uneven surfaces should be left to provide differences in water depth (Garbisch 1986; Leedy 1975). The vegetation zonation characteristic of most wetlands is due largely to the influence of water depth since different species are adapted to different depths of water (Hammer 1992). Where borrow pits are used, topsoil will need to be added as the bottom sediments of borrow pits are not likely fertile enough to encourage growth of wetland plants. Nearby wetlands may be used as biological benchmarks, with final depths in the created wetland designed to correspond with neighbouring ones (Garbisch 1986). When a project is an extension of an existing wetland, establishing similar grades on suitable soil should be sufficient to create proper hydrology.

There are a number of methods of revegetating new wetlands, including allowing natural germination to take place, seeding, planting of tubers or rhizomes, transplanting whole emergent plants, and planting trees from cuttings, dormant poles, or potted plants. While reliance on natural germination is less expensive, it is also much less reliable. Where possible, the upper 6-12 inches of soil from the original wetland should be transported (separately from the remaining soil profile) and used as the surface horizon for the created wetland. This will enable existing propagules in the soil to regenerate quickly (Lowry 1990). If soil from the original wetland is used, it should be stockpiled, kept moist (preferably underwater), and protected from leaching until it can be placed in the new site (Hammer 1992). Suitable conditions must be maintained to encourage natural invasion. For most species that requires moist, almost soupy muds, maintained by holding the water level at or immediately below the surface, or by periodic flooding and dewatering. It may take one to four years to achieve full vegetation coverage, but there is little control over what species eventually become established (Hammer 1992).

Transplanting vegetation from nearby wetlands may be a better option if plants can be taken from random, dispersed locations. Transplanting has the advantage of including the entire biological system associated with the roots (Lowry 1990).

It is advisable to attempt to provide uniform vegetative cover within one full growing season. Using a minimum of plant species adaptable to the various elevation zones of the site may be the most practical, allowing diversification to occur naturally (Garbisch 1986). If ground cover is not established by the end of the first growing season, exposed soil surfaces should be covered to minimize erosion.

To encourage development of a bog from a marsh, clumps of sphagnum and other mosses and acidophilic herbs, as well as shrubs and trees, should be distributed among marsh plantings (Hammer 1992).

Conclusion: Managing Wetland Impacts

Highway construction impacts on wetlands may be minimized by realigning the proposed routes to avoid wetlands, by using specific design and construction measures that mitigate potential impacts, or by replacing lost habitats with created or restored wetlands. The pros and cons of each of these three options need to be examined in detail to develop an appropriate management plan for wetlands affected by highway development.

The basic steps to wetland management for highways. A wetland management plan for highways has three basic steps as illustrated below. Each of these steps has associated guiding principles which help to identify key concerns. For example, avoidance, such as realigning the highway, generally is the first step taken to minimize impacts on wetlands. However, a decision to avoid a wetland requires recognition that there are links between terrestrial and wetland systems. Care should be taken to ensure that these linkages are examined before it is assumed that avoidance provides the remedy for all concerns. For example, placing a road adjacent to a wetland, rather than crossing it, may block an important wildlife migration route.

Mitigation perhaps is the most important step because highways cannot easily be made to twist and turn around each wetland. Mitigation during construction must attempt to preserve water flows, minimize surface disturbance, and provide needed wildlife migration paths to breeding sites in the wetland. It must be remembered that many amphibians which rely on wetlands for breeding spend most of their lives in terrestrial habitats. Flexibility in the design of flow structures should consider options for increasing or decreasing flows if monitoring indicates a need for water level manipulation.

Mitigation involves taking appropriate actions during construction, but also extends to monitoring and post-construction actions. For example, maintenance of flows may not be fully achieved by an initial design. Unforeseen design problems can be identified only through monitoring. Monitoring is best undertaken after sufficient time has elapsed for vegetation to show results of habitat change. This likely would take at least two years after construction.

Finally, replacement, wetland creation, should be part of any wetland management plan since the complexity of the wetland systems involved means that all concerns may not be adequately addressed through mitigation. Regardless of the care taken, some habitat loss likely will occur unless an effort is made to build new wetland habitat. At the same time, there is no certainty that a specific type of habitat necessarily can be constructed successfully. It makes the most sense to take advantage of opportunities that present themselves during construction. For example, small borrow pits may be left rather than filled, creating small ponds, or natural obstructions in drainage channels can be retained rather than removed. Taking advantage of opportunities likely will involve working closely with contractors in the field because many of the ideas will be opposite to common practice. The following chart illustrates and summarizes the steps and guiding principles required to practice sound wetland management for and during highway construction and maintenance.

Steps	**Guiding Principles**
(1) Avoidance	Recognize linkages between terrestrial and wetland systems
(2) Mitigation	Preserve water flows, minimize surface disturbance, provide migration paths, design in flexibility
(3) Replacement	Take advantage of opportunities

An important fact to remember while going through these steps is that it is extremely difficult to duplicate completely the characteristics and functions of a natural wetland. Therefore, wetland creation should be considered to be an additional compensation measure rather than a substitute for ensuring that existing wetlands are not harmed in any way as a result of highway construction projects.

References

Adamus, P.R., and L.T. Stockwell. 1983. *A method for wetland functional assessment.* Volume 1: *Critical review and evaluation concepts.* Federal Highway Administration, U.S. Department of Transportation, Washington, D.C.

Ducks Unlimited Canada. n.d. *Managing small wetlands for waterfowl.* Produced by LRIS, Amherst, Nova Scotia.

Erskine, A.J. 1992. *Atlas of breeding birds of the Maritime provinces.* Nimbus Publishing and the Nova Scotia Museum, Halifax.

Garbisch, E.W. 1986. *Highways and wetlands: compensating wetland losses.* Federal Highway Administration, Office of Implementation. McLean, Virginia .

Hammer, D.A. 1992. *Creating freshwater wetlands.* Lewis Publishers Inc. Boca Raton, Florida.

Halliday, T.R. 1993. Declining amphibians in Europe, with particular emphasis on the situation in Britain. *Environmental Reviews* 1(1):21-15.

Isnor, W. 1981. *Provisional notes on the rare and endangered plants and animals of Nova Scotia.* Curatorial 46. Nova Scotia Museum, Halifax .

Kusler, J.A., and M.E. Kentula. 1990. Executive summary. In *Wetland creation and restoration: the status of the science:*xvii-xxv. Edited by J.A. Kusler and M.E. Kentula. Island Press, Washington, D.C.

Leedy, D.L. 1975. *Highway-wildlife relationships.* Volume 1. *A state of the art report.* Federal Highway Administration, U.S. Department of Transportation, Washington, D.C.

Lowry, D.J. 1990. Restoration and creation of palustrine wetlands associated with riverine systems of the glaciated northeast. In *Wetland creation and restoration: the status of the science:*267-280. Edited by J.A. Kusler and M.E. Kentula. Island Press, Washington, D.C.

National Wetlands Working Group. 1988. *Wetlands of Canada.* Sustainable Development Branch, Environment Canada, Ecological Land Classification Series 24. Ottawa.

Zedlar, J.B., and M.W. Weller. 1990. Overview and future directions. In *Wetland creation and restoration: the status of the science:*405-413. Edited by J.B. Zedlar and M.E. Kentula. Island Press, Washington, D.C.

Progressive Reclamation Work At Cameco-Uranerz Key Lake Operations (1978-1995)

by D. Johannesen, L. Haji, and B. Zettl

Golder Associates Ltd., Cameco Corporation, and Prairie Plant Systems, Saskatoon, Saskatchewan

Abstract. The experimental regreening of disturbed sites at Key Lake, Saskatchewan started in the 1970s. Initially, numerous experimental plots were set up to study the behaviour of various agronomic species, their adaptability to the harsh environmental conditions that resemble sand dunes in the Athabasca region of northern Saskatchewan, and the best method of propagating them. The overall objectives of the program were to: reduce soil erosion, minimize dusting, rebuild the organic (top soil) layer, and improve the aesthetics of the site. However, the ultimate goal of any reclamation program is to return the disturbed land to a state similar to pre-development activities, in a self-sustaining condition. In 1993 Cameco-Uranerz augmented its reclamation efforts to include experimental programs designed to enhance the potential for long-term viability of revegetation programs. This process consisted of two parts: (1) the program design, and (2) experimental investigations. Prior to the initiation of the overall program, Cameco-Uranerz retained Golder Associates Ltd. to help design a study plan that would evaluate the success of different treatments in different substrates within the Key Lake mine site. The three main areas of investigations included: (1) stabilized hydroseeded areas, (2) sloped unstable areas (i.e. borrow pits and waste rock piles), and (3) exposed areas prone to wind/water erosion. Within each of these areas, various treatments with jack pine seedlings were employed to study their survivability. The primary concerns were moisture retention and wind protection. The experimental plots included: control, peat slurry dip, Alcasorb root dip, wood chip protection, shingle protection, desert tamers, and tire protection. The experimental plots were assessed in the spring and late fall of each year to determine the vigour of the seedlings. Based on the 1994 fall assessment, the survival rates ranged from 73% (shingle protection) to 89% (wood chip) and 71% (tire protection) to 94% (peat slurry dip) for the trees planted in 1993 and 1994 respectively. These recent studies indicated that the best method of reclaiming a disturbed site is to mimic the natural conditions prevalent in the area. To enhance this process the native or imported pioneer species capable of stabilizing a site and paving the way for climax species to survive, must be introduced. The new research also identified the limiting factor affecting the growth of climax species.

Introduction

The concept of ecological succession leading to a stable steady-state seral stage (climax) of a disturbed site, such as results from mining activities, has been well documented. In general there is a vast amount of literature available pertaining to reclamation of disturbed mine sites. Throughout the years, various ecological models such as Relay Floristics, Initial Floristic composition, Tolerance and Inhibition have been developed and tried (Winterhalder 1993). However, since the reclamation specialist has to deal with numerous and sometimes competing factors, he/she has to use experiments to test the nature of the site and what works best. In other words reclamation could be and often is very site-specific. To identify the most adaptable species and economical method of reclaiming the disturbed sites at Key Lake, numerous experimental plots between 1978 and 1986 were set up and studied. The present hydroseeding program is partially the result of these studies. However, due to competing priorities for limited manpower, not all of the results were properly documented. Also, at times many variables were changed at the same time, making it difficult to interpret the results. Therefore, in 1993, Cameco-Uranerz Key Lake Operation augmented its research program to include studying the pioneer and climax species and the relation between the two.

The new research program has two arms: (1) the laboratory program, where scientific research is performed on actual plants from the site, and (2) the field program, where various pioneer species and jack pine seedlings (the predominant climax species at the site) are studied under various conditions. The study started in 1993 and is ongoing.

This paper has two main parts: the first part briefly describes the initial study (1978-1986), and the second part describes the results and challenges of the field program, particularly the climax species, e.g. the jack pine seedlings. The paper also describes the factors contributing to the success of climax species. The results of pioneering species are still being compiled and no conclusive observation can be made at this time.

Background

The Key Lake Operation is a joint venture between Cameco Corporation and Uranerz. It is located in north-central Saskatchewan, about 70 km east-southeast of Cree Lake (Figure 1), at an approximate latitude of 57° 11′ north and longitude of 105° 34′ west (Figure 2). Geologically, it lies within the southern boundary of the Athabasca formation. Exploration of the area began in 1970 with aerial surveys, followed by reconnaissance with geological mapping commencing in 1971.

The discovery of a radiometric anomaly in 1971 and several more in 1972 assisted in locating the orebody. Drilling began in the winter of 1972-73 southwest of the orebody location with negative results. Further radiometric anomalies identified in 1973 assisted in focussing the search in the orebody area, with drilling finally intersecting the Gaertner orebody in the summer of 1975. Further exploration activities were undertaken at that time.

The project started in 1982 with the open pit mining of Gaertner ore body, which

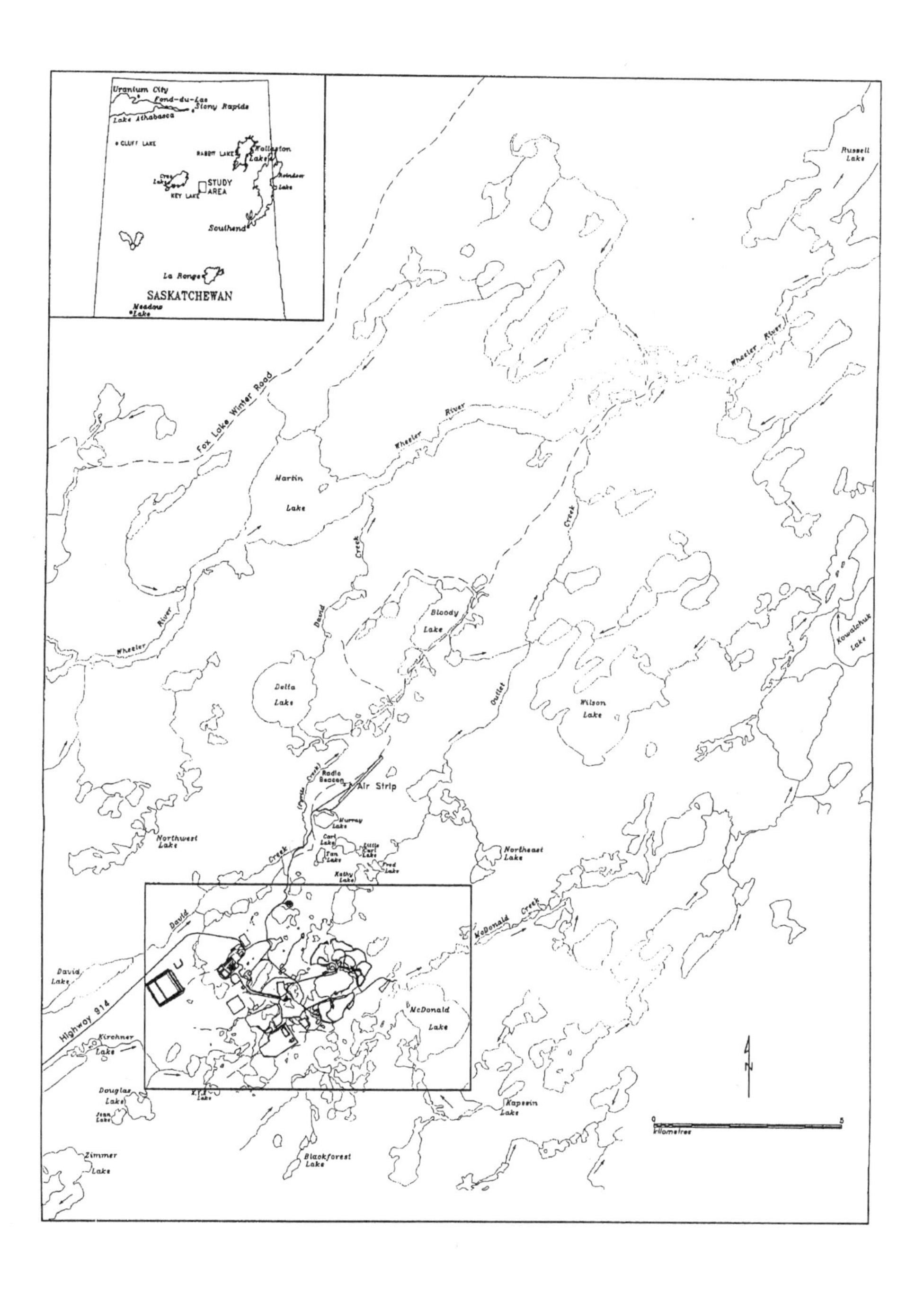

Figure 1. Study area and study location.

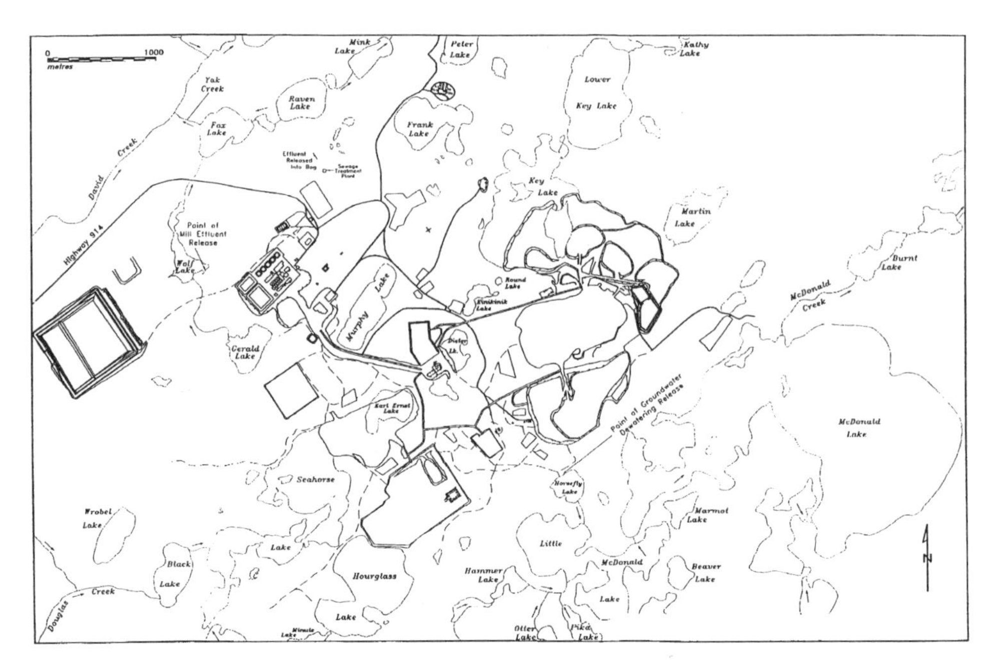

Figure 2. Project site map and location of facilities (inset from Figure 1).

was mined out in 1987. The open pit mining of Deilmann ore body commenced in 1986 and is expected to be mined out by 1997. The milling activities of the Deilmann ore are expected to continue for a few additional years following the completion of mining (TAEM & SENES 1994).

As a result of the mining activities, close to 1000 hectares of land have been disturbed which should be reclaimed. This includes the buildings sites, the tailings management facility, the reservoirs, monitoring ponds, the airport, the roadways, waste rock piles, laydown areas, the borrow pits, etc.

Environmental Factors Affecting Reclamation at Key Lake

Soil. There are five types of soil identified in the Key Lake area:

1. Till - composed of unsorted gravel, sand, and minor silt and clay glacially deposited as either ground moraine or shaped hills called drumlins.
2. Glaciofluvial - composed of mostly water-deposited sand and minor gravel, forming long sinuous ridges or eskers and also underlying most of the lakes, kames, irregularly shaped hills with steep slopes, and outwash sand plains.
3. Eolian Sand - restricted to small parabolic dunes in an isolated area (near Zimmer Lake).
4. Slope Wash - slope wash deposits consisting mostly of sand washed from hill sides to areas of lower elevation.
5. Organic Deposits - these are muskegs, present in the poorly drained areas and may reach a thickness of several metres. Muskegs without standing water contain permafrost.

Soils, where developed, are thin and underlain mostly by sand so that they are well-drained and exhibit low moisture-holding capacity. As a result, the Key Lake area is unsuitable for agriculture or pasture land.

All mineral soils near the Key Lake area have low nutrient status, Cation Exchange Capacity (CEC), pH, and electrical conductivity (EC) and are classified as sand textured. As a result, these materials are considered poor media for revegetation.

The organic soils have significantly higher nutrient contents, CEC, and moisture, but lower pH than mineral soils, and are considered better media for revegetation.

Climate and meteorology. The Key Lake area lies within the boreal climatic zone, which covers Canada from Alaska to Newfoundland. Boreal climates are influenced by cold arctic air masses causing long and frigid winters and cool, moist northern Pacific air masses which enter the region during generally short and cool summers.

Boreal climates are harshly cold with mean annual temperatures usually near or below freezing. Winters are long and cold while summers are short and cool. Rapid transition from winter to summer and from summer to winter cause short spring and fall seasons, and abrupt temperature changes are characteristics of these seasons.

The climate of the Key Lake area, therefore, is characterized by short, cool, moist summers with a mean monthly temperature of 15.7°C in the warmest month (July) and long, cold winters with a mean monthly temperature of -22.4°C in the coldest month (January). Extremes of mean monthly temperatures are typical of northern Saskatchewan.

The total annual precipitation has averaged 435.15 mm over the 1977 to 1994 period with 64.38% of the precipitation falling predominantly as rain over the summer months. The average pan evaporation rate over the same period was 669.03 mm (Annual Report 1994). The potential evapotranspiration (PE) is a measure of the amount of water that would be evaporated or transported from a surface completely covered with vegetation, if there was sufficient water in the soil at all time for use by vegetation. Hence PE is a better indicator of the runoff potential. The PE for Cree Lake, for which more comprehensive detailed information is available, is around 300 mm (based on 1969-1990 records). Based on data provided by Environment Canada's Geikie River stream discharge recording basin, which borders the Key Lake project area to the south, the average annual PE over the Geikie River drainage area was about 260 mm, representing approximately 45% of the average annual precipitation. As with temperature measurements, precipitation and pan evaporation rates at Key Lake are typical of those for northern Saskatchewan.

Average annual wind speed in the region is less than 16 km/hr, with maximum speeds between April and October, and relatively calm periods during winter which reduce the incidence of extreme windchill conditions. Winds from the west quadrant (SW-W-NW) and north quadrant (NW-N-NE) predominate (TAEM & SENES 1994).

Vegetation. Vegetative cover in the Key Lake area generally is a function of landform. There are two general vegetation types: forest type and wetland type. The forest type generally occupies a well-drained, treed upland site, whereas the wetland type occurs in poorly drained peat or sedge lowland sites.

The tree vegetation cover at Key Lake is almost entirely jack pine, with only small patches of black spruce. A forest fire in 1972 affected 40% to 50% of vegetation cover in the southern half of the site. Jack pine forests are found on upland areas with coarse textured soil. This includes both open jack pine forest with little ground cover (few or no shrubs, blueberry plants, lichen, etc.) and closed forests with dense shrubbery and forbs in the understorey. Most jack pine in the Key Lake area are from 85 to 150 years old (except in the burnt areas), and forest productivity and species diversity are lower than in other areas in the Athabasca basin.

Vegetation in lowlands with organic soil and poor drainage typically is dominated by open black spruce forest with an understorey of shrubs and mosses. Tamarack and swamp birch also are present in lowland areas, but are less common. Other predominant vegetative cover in wetlands are wet sedge meadow and rush-sedge cotton grass meadow.

Aquatic plant communities are common in slow sections of rivers and creeks in the Key Lake area. No plant species listed as rare or endangered in Saskatchewan have been found in the Key Lake area. Forest communities in the Key Lake area have no commercial value and no species in the area is utilized commercially at this time (TAEM & SENES 1994).

Reclamation and Revegetation (1976-1986)

The ultimate goal of reclamation at Key Lake is to restore the environment to a natural, self-supporting state with no further management input after mining and milling are complete. This means cleaning up or isolating any contaminated areas, allowing the ground and surface water regimes to recover, reclaiming waste storage areas, and restoring vegetation to the area. Therefore, reclamation may be quite different than revegetation. Revegetation, depending on the area, may be a part of the reclamation process or the reclamation itself. This report mostly deals with areas such that revegetation and reclamation share the same meaning.

Due to unforgiving environmental conditions, invasion of trees and shrubs in the Key Lake area is a very slow process. As a result, numerous studies were performed from 1978 to 1986 to identify the most effective:

- agronomic and native species;
- site preparation method;
- seeding methods and rates;
- fertilizer application methods, rates, timing and frequency;
- irrigation rates and methods;
- tree planting and maintenance procedures;
- soil amendments (lake sediments, lime, mulches, tackifiers, peat, hydrophillic agents); and
- transition from agronomic to self-sustaining vegetative cover.

Summary of results. Based on these studies the following conclusions were made:

1. The most successful species were: boreal red fescue, sheep's fescue, Carlton smooth brome, annual ryegrass and Fairway crested wheatgrass.
2. None of the legumes tested survived past the first year.
3. The native plants indicated low survival rates. In addition, they were found to be too costly and, therefore, not recommended for large-scale reforestation programs.

4. Containerized jack pine, if not fed upon by animals, survived well and are preferable to bare roots.
5. Seeds that are not covered by soil hardly ever germinate and are easily blown away or eaten by birds. The seed could be harrowed, raked, or covered by mulch. Seed application rates of 100 kg/ha were found to be the most effective.
6. Hydroseeding, despite its disadvantages, was found to be the most suitable and economic method for large-scale revegetation application.
7. The soil analysis indicated that the soil media are deficient in all major nutrients. An initial application of 100 kg/ha for each of N, P205, and K20 followed by an annual application of 100 kg/ha of N and P205 was proven to be very successful. Also, the timing of the application is very important (Courtney 1987).

As mentioned earlier, the results were not always documented properly and at times many variables were changed at the same time, making it difficult to interpret the data (TAEM & SENES 1994). As a result, in 1993 Cameco-Uranerz augmented its reclamation research program to include the pioneering and climax species prevalent at the site.

Ongoing Research

1993-94 research program. The new research program has two components: laboratory and field programs. The laboratory program is basically composed of identifying as many species as possible that grow naturally in the Key Lake area (both the pioneering and climax species), and collecting seeds, samples, or specimens from them and propagating the species in the nursery and, when ready, transplanting them in the field under various conditions. The key is to find the native species that can adapt themselves from ideal conditions in the nursery to the harsh conditions at the site. The field program is composed of planting the actual native plants grown under laboratory conditions, including the most prevalent species (e.g. jack pine seedlings), and studying the major factors affecting their growth and how well the plants can perform under various environmental conditions. These factors may include, but are not necessarily limited to:

- genetics of the species to be planted;
- soil quality of the area to be reclaimed;
- slope and aspect of the area;
- exposure to wind and the associated abrasion;
- ongoing erosion of the area; and
- moisture regime.

During the 1993 experimental season, the primary study areas were delineated between hydroseeded and non-hydroseeded areas of the mine site. This was to examine the potential competition between the grasses and the jack pine seedlings. The study sites were selected and the seedlings planted to the following treatments:

1. Control - Seedlings in this trial received no amendments, and were planted directly into the soil.

2. Peat Slurry Dip - Roots of each of the seedlings in this trial were dipped in a slurry of peat immediately prior to planting. The mixture is intended to provide a water-rich environment for the newly planted seedling, increased water retention capability of the soil, and a source of organic matter.

3. Alcasorb Root Dip - Each of the seedling's roots in this trial was dipped in a solution of Alcasorb and water prior to planting. Alcasorb aids seedlings by attracting water to the root system, thereby reducing dehydration.

4. Wood Chip Protection - Approximately one small pail (4 litres) of wood chips were spread out at the base of each seedling in this treatment. The chips were spread out around the seedling, extending out approximately 40 cm from the base of the seedling, with a 2.5 cm depth. The wood chips increase soil moisture retention and prevent grass encroachment. The chips also provide shade to the root collar of the seedling.

5. Shingle Protection - A pine shingle was placed approximately 15 cm from the seedling, on the south side of the seedling. These shingles were supposed to protect the seedlings from direct sunlight.

6. Desert Tamer - The latest version of the desert tamer was utilized for the 1994 experimental trials. Each desert tamer consisted of a hollow, rubber gasket and a one-piece sleeve that fit snugly into a 14-inch used rubber tire to create a reservoir for water storage.

7. Tire Protection - Where desert tamers were not available for the experimental trials, a single tire was placed around the seedling to create a protective microclimate.

The experimental design for 1993 consisted of a single treatment for each plot of seedlings. The design of the plots was somewhat different in 1994, with each plot composed of five to six rows, with each row receiving a different treatment. This change was made to avoid bias to any treatment, due to differences in soil quality, exposure, or moisture regime at a given site.

Fertilizer stakes used in 1993 were omitted from the 1994 study, as they were noted to "burn out" the jack pine seedlings and enhance competitive grass growth around the trees that did survive. The other amendment that was deleted from the 1994 study was shingles placed on the south side of the seedlings. Although this was meant to shade the area to slow down moisture loss, the trees were noted to be growing to the side, apparently attempting to get into direct sunlight.

The sites containing the 1993-1994 experimental plots are described below:

Site A - The "old construction camp" site which has a slight south-facing aspect and is located near the Key Lake entry gate. The site was successfully hydroseeded approximately four to five years ago and as a result has good grass cover. This area also has been fertilized on an annual basis. The main grass species that germinated are boreal red fescue and sheep's fescue. Some smooth brome and crested wheatgrass also can be found scattered throughout the site. In the low areas on the south-east end of the study site, willows have pioneered naturally, suggesting a high water table. Good soil moisture conditions were present during the time of planting. The site was planted with 1,650 jack pine seedlings in 1993, and with 1,750 in 1994.

Site B - The area is a low, tailings pit borrow area. It is mostly bare with a few natural revegetating plants (willow, balsam poplar, grass species). Because of the low elevation of the site, soil moisture has been very high, with some standing water in rutted areas throughout the site. Some jack pine have started to revegetate a south-facing slope at the northwest corner of the site. In 1993, 2100 seedlings were planted and 1382 seedlings were planted in 1994.

Site C - The disturbed site consists of approximately 33 acres of cleared jack pine. It has a slight south aspect and is surrounded by regenerating jack pine. The site was hydroseeded in 1992, resulting in a good mulch cover with approximately 50% grass germination at the time of tree planting. The mulch cover has resulted in good moisture retention conditions in the sand. A total of 1050 seedlings were planted in four plots on this site in 1993. In 1994, 3500 more seedlings were planted. Of these seedlings, 500 were planted in an intensive (5 m x 4 m) plot, with a 25 cm spacing.

Site D - Four hundred trees were planted along an existing cleared trail through a jack pine area. The trail was approximately 10 m wide and ran from an existing road in a northwest to southeast direction. Four rows of 100 trees each were planted at 2 m intervals. Rows of "new stock" and original stock alternated from west to east. Due to insufficient space, no trees were planted at this site during the 1994 program.

Site E - Two hundred trees were planted in 1993, in three rows in an open sand area adjacent to an existing road. The rows consisted of 70 "new stock" trees, 60 original stock trees and 70 "new stock" trees. Rows ran in a northwest to southeast direction. In 1994, 2150 trees were planted, of which 50 used desert tamers and 500 were placed in an intensive plot similar to the ones at Site C.

Site F - This site was established in 1994. It is a sandy, southeast facing sloped area, located directly east of Site E. A total of 600 trees were planted at this site, including 76 desert tamers.

Site G - Also established in 1994, this site contains three plots on a northwest facing, steep slope area located south of the crusher. The plots contain a total of 800 seedlings.

Each plot on all of the sites was visited and the individual trees assessed. The trees were rated from 1 to 4, as defined below:

1. Good - seedling is green with no signs of dead or brown needles.
2. Average - seedling is greater than 50% green.
3. Poor - seedling is greater than 50% brown (winterkill); buds green and actively growing.
4. Dead - seedling dead or missing (seedlings uprooted by animals).

In order to monitor the success of the 1993 and 1994 planting programs, the experimental plots were evaluated using established protocols during the spring of 1995. The results are shown in Tables 1 and 2. Overall the 1993 and 1994 seedling survival rates were 79.5% and 91.1% respectively, at the time of the spring 1995 survey.

1995 research program. Another important factor in the identification of suitable species for reclamation projects in Saskatchewan has become the avoidance of introducing competitive non-native species into the ecosystem. Typically, these are aggressive species (e.g. crested wheatgrass), and they have been well suited to reclamation as they grow quickly and provide suitable erosion protection. However, they have also out-competed native species in the southern regions of the province and in places have created monocultures that are less appealing to wildlife.

In order to address the utilization of native vegetation species for reclamation at Key Lake, several species were collected (e.g. plants and cuttings) from northern Saskatchewan and taken back to the greenhouse for propagation. Species currently under investigation on the Key Lake site include:

Grasses

Calamagrostis neglecta - narrow reed grass
Carex foenea - hay sedge
Bromus pumpellianus - northern awnless brome
Festuca rubra - red fescue
Calamagrostis inexpensis - northern reed grass
Agrostis scabra - rough hair grass

Woody species

Vaccinium sp. - blueberry
Salix sp. - willow
Populus balsamifera - balsam poplar
Populus tremuloides - trembling aspen
Chamaedaphne calyculata - leatherleaf
Betula sp. - birch
Prunus pennsylvanica - pincherry
Pinus banksiana - jack pine

In order to further test certain species success rates for survival and vigour under differing habitat conditions within the Key Lake lease area, Golder Associates and Prairie Plant Systems designed and established several test plots which were planted to the above species.

Three sites were chosen that represented the environmental conditions throughout the mine site. At each of these sites there was an experimental plot design with 16 plots representing four replicates of each of the following treatments:

T1 - control with no wind fence and no hydromulch and fertilization

T2 - wind protection (snowfence) with no hydromulch

T3 - wind protection (snowfence) with hydromulch

T4 - no wind protection (exposed) with hydromulch

Each of the plots that was planted with grass plugs around the perimeter was planted with 70 jack pine seedlings (7 rows of ten seedlings each). The experimental sites are described below:

Site 3 - Moderately harsh environment - exposed area, south of the mine shop, comprised of loose sand hauled onto a black spruce bog. The sand thickness is approximately 2 m on the north portion of Site 3 where the plantings were situated.

Site 17 - Mild environment - this is a protected area along the road between the mine shop and camp. The area has some residual vegetation, particularly along the northern and southern boundaries and is surrounded by mature and regenerating vegetation.

Site 21 - Extremely harsh environment - this is a south-facing slope of the waste rock pile. The main component is sand with some larger rock present. The slope was modified to a 3.5:1 grade.

The success of each species discussed above is currently being assessed.

Additional experimental seedling plots, 1995. In addition to the grass experimental plots, other plots were established in areas around the lease to evaluate different treatments on seedling vigour and survival.

Table 1. Key Lake Jack Pine Seedling Assessment - 1993 Plantings (July 1995)

Plot	Number of Trees	Treatment	Survival Assessment (%)[1]			
			Good (1)	Average (2)	Poor (3)	Dead (4)
A-01	300	Wood Chips	28	63	1	8
A-02	300	Shingles	13	77	4	7
A-03	300	Root Dip	8	37	5	50
A-04	300	Peat	8	62	6	24
A-05	300	Control	21	58	8	13
A-06	106	Trees Only	24	62	7	8
A-06	44	Tires	25	39	0	36
Mean			18.0	56.8	4.3	20.8
B-01	300	Wood Chips	19	69	2	10
B-02	300	Shingles	13	41	5	41
B-03	300	Root Dip	38	55	1	7
B-04	300	Peat	11	65	9	15
B-05	300	Control	37	56	3	5
B-05	50	Wood Chips	2	82	2	14
B-06	150	Desert Tamer	29	57	1	14
B-06	150	Tires	24	73	1	2
B-07	300	Slope	6	19	5	69
Mean			19.8	57.5	3.1	19.6
C-01	300	Peat	17	74	3	7
C-02	300	Control	12	78	3	7
C-03	150	Desert Tamer	20	67	0	13
C-04	300	Vegetation	13	79	3	5
Mean			15.6	74.2	2.1	8.1
D-01	100	Old	4	40	9	47
D-02	100	New	27	60	0	13
D-03	100	Old	14	40	12	34
D-04	100	New	12	70	11	7
Mean			14.3	52.5	8.0	25.3
E-01	70	New	4	80	16	0
E-01	60	Old	2	50	18	30
E-01	70	New	7	86	7	0
Mean			4.4	71.9	13.7	10.0
Overall Survival Mean Percentages			15.2	59.0	5.3	17.1

[1] Due to rounding, percentages may not total 100.

Table 2. Key Lake Jack Pine Seedling Assessment - 1994 Plantings (July 1995)						
Plot	No. of Trees	Treatment	Survival Assessment (%)[1]			
			Good (1)	Average (2)	Poor (3)	Dead (4)
94-A	350	Control	20	75	6	2
94-A	350	Peat	16	66	16	1
94-A	350	Root Dip	28	68	3	1
94-A	350	Shingles	29	48	14	0
94-A	350	Wood Chips	29	67	1	3
Mean			24.4	64.8	8.0	1.4
94-B	250	Control	6	90	2	2
94-B	250	Peat	10	87	0	3
94-B	282	Root Dip	19	87	1	6
94-B	250	Shingles	11	84	2	3
94-B	250	Wood Chips	24	68	2	6
Mean			14.0	83.2	1.4	4.0
94-C	600	Control	8	79	4	9
94-C	600	Peat	7	83	3	7
94-C	600	Root Dip	7	84	1	8
94-C	600	Shingles	5	73	3	20
94-C	600	Wood Chips	4	75	7	14
Mean			6.2	78.8	3.6	11.6
94-E	200	Control	1	75	7	18
94-E	50	Desert Tamer	0	86	2	12
94-E	200	Peat	1	77	14	9
94-E	200	Root Dip	2	71	12	16
94-E	200	Shingles	2	68	12	19
94-E	100	Tires	9	60	2	29
94-E	200	Wood Chips	2	47	34	18
Mean			2.4	69.1	11.9	17.3
94-F	100	Control	2	43	51	4
94-F	76+24 Tires	Desert Tamer	25	69	4	2
94-F	100	Peat	3	39	52	6
94-F	100	Root Dip	5	37	46	12
94-F	100	Shingles	0	66	20	14
94-F	100	Tires	3	42	42	13
Mean			6.3	49.3	35.8	8.5
94-G	150	Control	11	80	2	7
94-G	50	Desert Tamer	42	42	0	16
94-G	150	Peat	13	81	1	5
94-G	150	Root Dip	1	80	9	9
94-G	150	Shingles	10	84	1	5
94-G	150	Wood Chips	12	72	3	13
Mean			14.8	73.2	2.7	9.2
Overall Survival Mean Percentages			10.7	69.2	11.2	9.2

[1]Due to rounding, percentages may not total 100.

Site 4 - Mill Hill area directly between the mill and the mine shop. Four 10 m x 10 m plots were established; two on the crest of the hill and two on the side-slope facing the mine shop. At each location, there is a control plot and a plot that was treated with approximately 6.25 kg of lime. Each plot then was planted with 100 jack pine seedlings for a total of 400 seedlings.

Site 21 - Treatments that were tried at this location include slope modification and seedling protection. These plots were located west of the grass experimental plots and consisted of the following:

1. Small shallow pits (shelter pits), approximately 20 cm wide and 10 cm deep were dug, with the soil from the hole placed on the north side of the seedling. The intent of this treatment was to create a small wind protection mound for the seedling, while creating a depression to hold moisture. The seedling was planted half way up the created slope so that it was protected by the mound, but would not be flooded.
2. A portion of the side-slope of Site 21 was delineated into four, 100 m sections. Two sections were altered by walking a bulldozer perpendicular to the slope to create a series of ridges on the slope-face. The intent was to create a more protective, stable environment for the seedlings to become established, particularly during the first year. The ridges could decrease the erosive force of the wind, protect the seedlings from the abrasive blowing sands, aid in moisture retention, and decrease erosion caused by runoff. The other two sections were left untreated, with respect to terrain alteration.
3. Half of the control and the ridged areas were covered with mulch to determine the effects of the added erosion control.

All four areas then were planted with jack pine seedlings. Due to the uneven ridges, the numbers of seedlings in each area were not identical.

Site E - This site is located immediately north of the Gaertner Pit and is exposed, with little surrounding cover. Due to this exposure plants in this area are subjected to "sand blasting." In addition, any seeds are blown away from the area so that seed germination of any kind is prohibited. In Site E several treatments were established in an attempt to protect the seedlings from the wind and blowing sand. The treatments were: pits, control, shingles, Alcasorb, and windfence. The windfence was an area approximately 12 m x 12 m surrounded by orange plastic windfence. The windfence is porous, and will decrease the wind speed, thereby minimizing the abrasive effects of the sand. The fence also may enhance moisture retention of the soil by decreasing evaporative potential of the wind. This will allow the seedlings to establish themselves better.

Grass seed mixture trials. At Sites 3, 17, and 21 grass seed mixture trials were conducted to determine the success of commercially available cultivars, including: Durar hard fescue, Streambank wheatgrass, slender wheatgrass, Canada blue grass, northern wheatgrass, and creeping red fescue. The areas were fertilized at a rate of approximately 27.5 kg/ha of nitrogen and 5.7 kg/ha of potassium. The seeding rate was approximately 1.5 kg/ha, with the seeds broadcast and lightly raked. Each plot was 2 m x 2 m.

At each of the three sites, one series of plots was covered with hydromulch and the others were used as a control.

Seedling development experimental area. This plot was established in Site C for the measurement of root and stem development of the seedlings planted in 1995. Every

Table 3. Summary Gain in Shoot, Root, and Biomass (Jack pine seedlings 1995)							
Length (cm)							
Shoot[1]				Root[2]			
Greenhouse	July 11	August 23	Gain	Greenhouse	July 11	August 23	Gain
20.40	20.51	22.35	1.95	—	10.83	18.15	7.32
Biomass[2] (g)							
Shoot				Root			
Greenhouse	July 11	August 23	Gain	Greenhouse	July 11	August 23	Gain
—	0.99	2.63	1.64	—	1.31	2.03	0.73

[1]Gain is the difference between the greenhouse and August 23 values.
[2]Gain is the difference between the July 11 and August 23 values
— = Not measured
Note: All values are in mean number of trees:
Greenhouse=380 seedlings
July 11=10 seedlings
August 23=10 seedlings

four to six weeks, 10 seedlings will be collected and the length and dry weight of the stem and roots of the seedling will be measured. This will allow us to evaluate the development of the seedlings and to determine the shoot:root weight ratio for the seedlings.

The results of the seedlings measured on July 12, 1995 as compared to measurements on August 23, 1995 are shown in Table 3. The measurements were taken from collar to shoot tip and collar to root tip. The samples were cleaned of their rooting medium and dried for 48 hours at 50°C. Care was taken not to remove fine rootlets. After the samples were dried, their weights were determined with a Mettler electronic balance at the Key Lake chemistry laboratory.

Conclusions

To date, the survival rates for the jack pine seedlings for the 1993 and 1994 plantings are 79.5% and 91.1% respectively. This includes all treatments on all study areas.

1993 Plantings. On Site A (hydroseeded), the shingles, wood chips, and trees only (control) treatment seedlings exhibited the highest survival rates at 93%, 92%, and 92% respectively. The only attributable difference in these treatments was that the wood chips were acting to decrease the competition from the grasses.

On Site B (not hydroseeded), tires (98% survival), control (95% survival), and root dip (93%) exhibited the best seedling performance. The tires provided a microenvironment for the seedlings that was conducive to growth (i.e. moisture retention), and also provided protection from the abrasive sands.

On Sites D and E the seedling survival rates illustrate the importance of seedling vigour at time of planting. The old stock seedlings were in poor condition and stressed at the time of planting. New stock seedlings were transported to site and immediately planted beside the old stock. The new stock was moist and unstressed, and apparently was better able to cope with the additional stress of planting.

1994 Plantings. The differences in seedling survival between treatments of the 1994 plantings on Sites A and B are not immediately obvious. However, the wood chips and root dip treatments do show higher percentages of seedlings in good shape on these two sites.

On Site C, the peat (93%), root dip (92%), and control (91%) treatments exhibited the highest survival rates for seedlings, but again the differences between all treatments were slight.

The desert tamer, peat, and control treatments were the top three on Sites E and F. Site F was a sandy, exposed slope with a southeast aspect. Under this condition, the added moisture provided by the desert tamer appeared to be a factor in the survival and vigour of the seedlings (i.e. higher number of seedlings in good condition).

This same trend was noted for Site G, a steep (25%) slope with a west-facing aspect, where the desert tamers provided the highest number of seedlings in good shape. Although the desert tamers provided higher percentages of Class 1 seedlings on steep slope areas, their use at Key Lake has been eliminated from consideration for large-scale use. This is due to the high installation costs, and the damage to the soils and seedling root system when the unit is removed. During final reclamation plans, the desert tamers may be used for site-specific areas to increase the viability of trees on a small area.

Although there are subtle differences in performance of jack pine seedlings under the various treatments, the only clear trend that was noted was that the control seedlings generally appeared in the top three treatments. Long-term data will need to be collected before any definitive statements on the effects of the treatments can be made.

Through direct observations on site, the primary obstruction to reclamation on the mine site appears to be wind and water erosion. The blowing sand "sand blasts" the seedlings to the point where needles are literally knocked off. On steep slope areas, water erosion from spring melt and heavy summer rains causes significant damage to unprotected sloped areas.

These factors have been recognized and are being studied on site. Currently, the benefits of protection measures such as ridging the slope surfaces, covering slopes with hydromulch, using snowfence, and tree pits are being explored.

References

Cameco Corporation. 1994. Key Lake operation. Environmental Annual Report.

Courtney, P. 1978 - 1987. 1987. Revegetation Cameco Corporation Internal Document.

Testwork at Key Lake:
Golder Associates Ltd. 1995. Progress report for the Key Lake regreening project.

Prepared for Cameco Corporation:
TAEM & SENES Consultants. 1994. Status of environment report. Cameco Corporation, Key Lake Operation.

Winterhalder, K. 1993. *The roles of colonization and succession in reclamation of mine sites.* 10th National Meeting of American Society for Surface Mining and Reclamation. Spokane, Washington. May 16-19, 1993.

An Ecological Land Classification Based Land Management Plan for Old Mined Lands

by Trent Enzsol

Environmental Planner, Environmental Services Department, Prairie Coal Ltd., Estevan, Saskatchewan S4A 2K9

Abstract. Prairie Coal Ltd. controls approximately 2500 hectares of land near Estevan, Saskatchewan which were mined between the early 1930s and early 1970s, prior to the implementation of formal reclamation guidelines. In June 1993, Saskatchewan Environment and Resource Management approached Prairie Coal Ltd. to develop an environmentally based plan to manage these lands. The land consists of a series of overburden ridges, referred to as spoil piles, and numerous ponds in old box cuts and end pits. Prairie Coal Ltd. developed a Land Management Plan using an Ecological Land Classification (ELC) approach. The ELC entailed classifying recurring abiotic and biotic factors into similar ecological land units. The old mined lands fell into four ecosites: Type 1, with slopes less than 10°; Type 2, with slopes between 10° and 25°; Type 3, with slopes greater than 25°; and End Pits/Box Cuts, with steep banks. As slope angle on the spoil piles increases, spoil height, salinity, and sodicity increase and vegetation cover decreases. Based on the dominant vegetation observed, the ecosites were delineated into 11 ecoelements. The ELC data were used to assess the habitat suitability of the different ecosites for mule deer (*Odocoileus hemionus*) and Sharp-tailed Grouse (*Pedioecetes phasianellus*) using Habitat Suitability Index models (HSI). The HSI models ranked the habitat suitability of Type 2 > Type 1 > Type 3 > End Pits/Box Cuts. Seasonal utilization surveys showed mule deer using Type 2 and 3 spoils more than Type 1 spoils and End Pits/Box Cuts. The ELC and wildlife habitat assessment provided an ecologically defendable justification for wildlife habitat/pasture as the best end land use for the Land Management Plan without conducting large-scale reclamation work. The Plan identified aesthetic concerns along major travel corridors where partial levelling and revegetation trials have been initiated. The ELC served to select seed mixes, tree and shrub species, and resloping angles to be used for these trials.

Introduction

Prairie Coal Ltd. (PCL), a subsidiary of Manalta Coal Ltd., controls approximately 2500 hectares of old mined land located in the southeastern part of Saskatchewan near the City of Estevan and the Souris River (Figure 1). These lands were mined between the early 1930s and early 1970s, prior to the implementation of formal reclamation

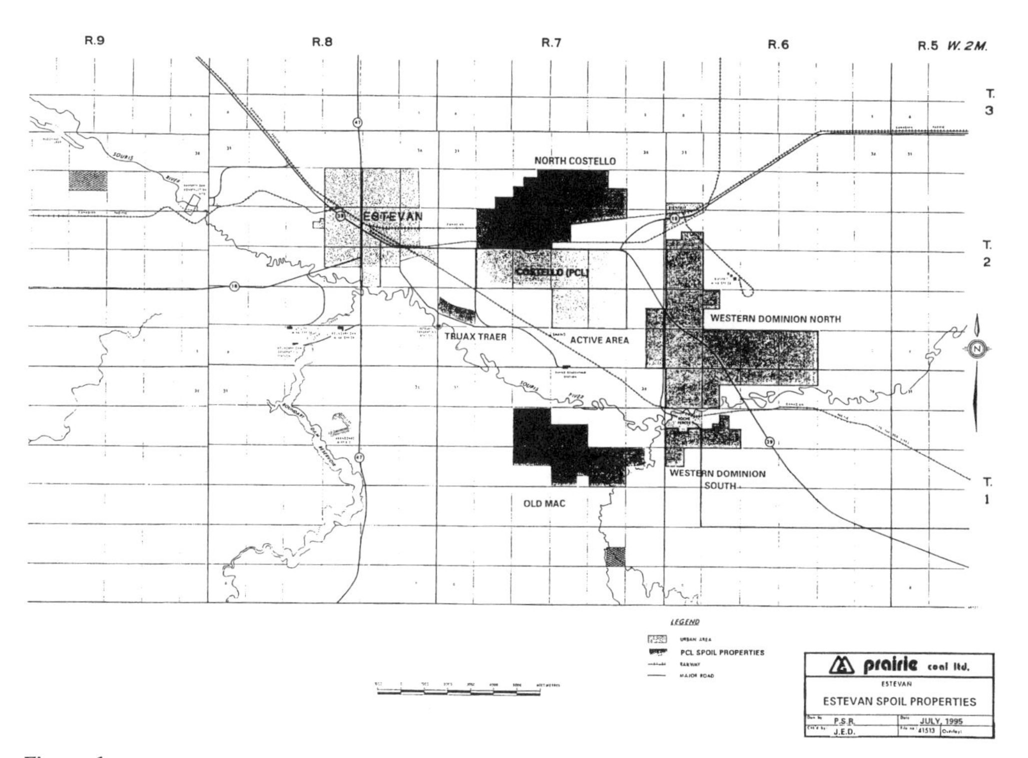

Figure 1.

guidelines. The old mined land consists of a series of overburden ridges, referred to as spoil piles, and numerous ponds in old box cuts and end pits.

In 1964, predecessors of PCL acquired the Costello Mine (formerly Klimax Mine) which included several previously mined properties consisting of the Old Mac, Truax Traer, and Western Dominion strip mines. PCL refers to these properties as the "Estevan Spoil Properties." Since the early 1930s, various large-scale surface mining techniques, including horse and tractor drawn scraper operations, shovel operations, and dragline operations, have been used to strip off the overburden material to expose underlying coal seams. Because of the differences in these mining techniques, the age and variations in overburden material and the vegetation cover differ considerably throughout the old mined lands.

Unlike today, these properties were mined with no reclamation guidelines in place. At the time, reclamation plans and end land use plans were not developed, valuable topsoil was not salvaged, and spoil piles were not levelled. This contributed to the delivery of a continuous and economical supply of coal for domestic and commercial fuel for the Estevan area. By 1971, the first reclamation requirements were introduced by the Saskatchewan government. These requirements stated that "all operations involved in strip mining coal must submit reclamation plans to the Government." The early reclamation plans usually consisted of levelling spoil piles and revegetating with a grass/legume seed mixture. In 1984, these early reclamation requirements were replaced when Saskatchewan Environment and Public Safety implemented the "Reclamation Guidelines for the Estevan Mining Area." To meet the terms of these guidelines, mining operations were required to submit detailed reclamation plans which outlined cover soil salvage, revegetation measures, and end land use plans. Currently coal mines in southern Saskatchewan operate under the "Mineral Industry Environmental Protection Regulations" (1991) and the updated "Reclamation and Licencing Guidelines" (1993).

This paper outlines an Ecological Land Classification (ELC) approach for describing old mined lands. The objectives were to develop a management plan for the Estevan spoil properties and to guide selective partial levelling and revegetation programs. Wildlife study techniques are also utilized to guide planning. Results relative to these programs are summarized.

Project History

In 1993, several meetings were held between Saskatchewan Environment and Resource Management (SERM) and other corporations involved with mining the Estevan area. The purpose of these meetings was to identify land management issues associated with the old mined lands, and to agree upon a course of action to ensure that the lands are in a condition suitable for release from further decommissioning and reclamation obligations from SERM. During these meetings, the following observations regarding the old mined lands were made:

- The spoil piles were created during a time when there were no reclamation guidelines in place.
- Many of the areas have revegetated successfully naturally and support a variety of wildlife uses and pasture opportunities. Little would be gained by large-scale levelling of these spoil piles because natural vegetation communities would take a long time to re-establish.
- Some of the spoil piles may present aesthetic concerns, particularly those which are poorly vegetated and located along major corridors like Highway 39.
- Due to mining equipment limitations of the day, the deeper coal seams remain beneath the old mined properties; therefore, there is potential to re-mine these areas if there is the demand for it.
- Those spoil piles which have been partially reclaimed and those that have naturally revegetated likely have achieved a useful end land use condition as pasture/wildlife habitat.

It was agreed by all parties that a Long-Range Spoils Management Plan be developed by the mining companies for the old mined lands they controlled.

Methodology for the Development of a Long-Range Spoils Management Plan

PCL decided that an ecological approach to developing a land management plan for the old mined properties which could verify a wildlife habitat/pasture end land use would be appropriate for the situation. To develop an ecologically based land management plan, PCL formulated an ecological land classification (ELC) and conducted a wildlife habitat assessment. This work was initiated in 1993 and carried out by staff of PCL's Environmental Services Department in consultation with SERM.

The study area consists of the old mined areas of Old Mac, Western Dominion, and North Costello mines (Figure 1).

Ecological land classification. The objective of an ecological land classification (ELC) is to survey, classify, and map recurring biotic and abiotic factors into similar ecological units. This method of land classification entails description, comparison, and synthesis of data related to the biological and physical characteristics of the land unified into one, rather than separate maps of soil, geology, climate, topography, and vegetation (Rowe 1977).

The ELC framework describes the Estevan properties according to the five hierarchal levels below:

- Ecoregion: an area characterized by a distinctive regional climate as expressed by vegetation (Wilken and Ironside 1977);
- Ecodistrict: subdivisions of an ecoregion which are characterized by a distinctive pattern in relief, geology, vegetation, and geomorphology (Strong 1979);
- Ecosection: subdivisions of ecodistricts which are determined by recurring patterns or assemblages of local landforms, topography, soils, and vegetation (Strong 1979);
- Ecosite: subdivisions of ecosections which are based on specific landform units, topographic situation, and climax or dominant vegetation type (Wilken and Ironside 1977);
- Ecoelement: subdivisions of ecosites which are determined by the dominant or successional stage of vegetation (Wilken and Ironside 1977).

The ELC consisted of three phases:

1. Pre-field literature review and preliminary mapping.
2. Field investigations:
   - inventory of vegetation, chemical and physical soil characteristics, spoil slope, and height, and
   - inventory of present land uses and verification of preliminary ELC mapping.
3. Post-field analysis and ELC map production.

The ELC mapping was completed in 1994.

Wildlife habitat assessment. PCL used Habitat Suitability Index (HSI) models to conduct its wildlife habitat assessment. HSI models serve to evaluate the habitat potential of a habitat type assuming selected measurable biophysical variables govern the ability of a particular habitat to provide the life requisites for a selected species. The habitat suitability index is determined by using habitat suitability index graphs for specific variables which range from 0 to 1.0. Suitability index values of 0 represent no suitable habitat while those at 1.0 represent optimal habitat. For the old mined land two wildlife species were selected for the evaluation: mule deer (*Odocoileus hemionus*) and Sharp-tailed Grouse (*Pedioecetus phasianellus*). These two species were selected because of their recreational importance in the area.

An ELC approach provides a good basis for implementing HSI models since the

ELC considers a variety of factors which reflect and influence wildlife behaviour (Demarchi 1985). These factors relate to climate, moisture, slope, aspect, geology, glaciation, and plant community structure (Demarchi 1985). To implement the models the biophysical variables associated with the mule deer and Sharp-tailed Grouse models were identified from the models and measured in the field. Most of the variables required to run the models were measured for conducting the ELC. The habitat suitability index was calculated at the ecosite level.

Summer pellet group counts and winter aerial surveys of deer were conducted to estimate the seasonal utilization of the various ecosites by deer. The adjacent upland, coulee, and floodplain areas were included in the winter aerial surveys.

Results

The ecological land classification for the old mined properties consisted of one ecoregion, one ecodistrict, one ecosection, four ecosites, and eleven ecoelements. Figure 2 schematically presents the results of the ELC.

Ecoregion and ecodistrict. The old mined land lies within the Moist Mixed Grassland ecoregion and Saskatchewan Plains ecodistrict. The Moist Mixed Grassland and Saskatchewan Plains are correlated closely with semi-arid moisture conditions and dark brown Chernozemic soils. Most of the ecoregion and ecodistrict are characterized by hummocky to gently undulating glacial till plain.

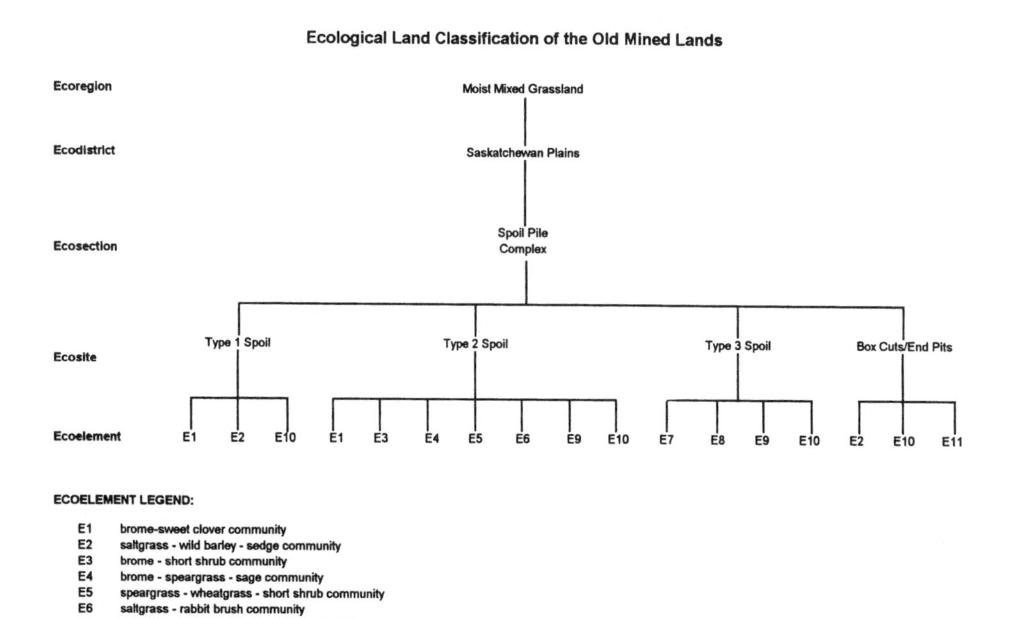

Figure 2.

The Moist Mixed Grassland and Saskatchewan Plains are dominated by two vegetation community types:

- speargrass-wheatgrass community (*Stipa* spp.-*Agropyron* spp.); and
- speargrass-wheatgrass-blue grama grass community (*Stipa* spp.-*Agropyron* spp.-*Bouteloua gracilis*).

During drought years and/or over-grazing the invasion of blue grama grass is favoured, which causes a shift in species composition from that of the more typical speargrass-wheatgrass community to one of a speargrass-wheatgrass-blue grama grass community. In years of improved moisture conditions and reduced grazing pressure, recovery towards the more typical speargrass-wheatgrass community is expected (Abouguendia 1990).

Ecosection. The old mined lands were classified into the Spoil Pile Complex ecosection. The Spoil Pile Complex is characterized by a series of overburden ridges, termed spoil piles, and numerous wetlands in old box cuts and end pits. The mining disturbances created between the early 1930s and early 1970s have changed the landscape to the extent that the soil and topographic conditions have resulted in plant communities distinct from the adjacent till plain, coulees, and floodplain.

The topographic variability of the Spoil Pile Complex ranges from less than 3 m high to greater than 15 m high, with slopes from less than 10° to greater than 35°. Most of this topographic variability can be attributed to the method of mining utilized to strip mine the land.

The soils of the Spoil Pile Complexes cannot be described in the usual pedological meaning of the term; rather, they are a variable mixture of topsoil, glacial till, and bedrock. The soil descriptions in this report refer to the top 50 cm of the spoil pile, also referred to as the rooting zone. The composition of the rooting zone varies from:

- predominantly glacial till mixture;
- predominantly glacial till/bedrock mixture; and
- predominantly bedrock mixture.

In general, as bedrock content increases salinity and sodicity increase and the texture changes from a sandy loam to clay loam. The soil composition and quality reflects the method of mining used in a particular area. During earlier mining operations, horse and tractor drawn scrapers, shovels, and small draglines did not have the ability to make deep cuts in the overburden; therefore, the spoil piles created are predominantly a glacial till mixture or a glacial till/bedrock mixture. During later mining operations, larger, more modern, draglines had the ability of making deep cuts in the overburden; therefore, these spoil piles are predominantly a bedrock mixture. Spoil piles which are contaminated with saline and sodic bedrock are less conducive to revegetation cover.

The most common species of vegetation observed along the ridges and gullies within the Spoil Pile Complex include brome grass (*Bromus inermis*), wheatgrass (*Agropyron* spp.), speargrass (*Stipa* spp.), salt grass (*Distichlis stricta*), sage (*Artemisia* spp.), sweet clover (*Melilotus* spp.), caragana (*Caragana arborescens*), Russian olive (*Elaeagnus angustifolia*), snowberry (*Symphoricarpos occidentalis*), and rose (*Rosa* spp.). In general, the higher, steeper slopes of the Spoil Pile Complex have less vegetation cover than the shorter, gentler sloped piles because of the steep spoil slopes and saline and sodic soil conditions of the rooting zone.

The strip mining that created the Spoil Pile Complex also resulted in a number of box cuts, end pits, and valleys between spoil piles which contain varying amounts of water. Common species of vegetation found in the wetland areas include cattails (*Typha* spp.), sedge (*Carex* spp.), bulrush (*Scirpus* spp.), rush (*Juncus* spp.), wild barley (*Hordeum jubatum*) and wheatgrass.

Land uses within the Spoil Pile Complex are influenced by the topographic limitations. Cereal and forage crop production is limited by the soil quality and topography. The primary land uses are restricted to grazing and wildlife habitat. A number of the ponds maintain a water depth that can support fish populations. Fish species which have been stocked successfully include rainbow trout, pike, perch, splake, and walleye.

Ecosites. The Spoil Pile Complex was delineated into four ecosites: Type 1, 2, and 3 spoils, and Box Cuts/End Pits. These ecosites were delineated based on spoil pile slope and height. The spoil pile height, in general, indicates the soil quality, plant community, and revegetation success. Figure 3 is an example of the ELC mapping at the ecosite level for the Western Dominion area.

Ecosite 1: Type 1 Spoil. The Type 1 spoil ecosites are the result of strip mining with horse or tractor drawn scrapers, small mining shovels, or partially levelled reclaimed land. The Type 1 spoil ecosites are on average less than 3 m high and have slopes less than 10°. These areas are among the oldest spoil piles in the Estevan area.

The soil conditions of the Type 1 ecosite vary from having extreme limitations to minimal limitations for plant growth. Areas with vegetation cover less than 30% are characterized by highly saline and/or sodic soil conditions. The mean (±SD) Sodium Adsorption Ratio (SAR) of the top 20 cm is 26.7 ± 7.8 and from 20-50 cm is 25.6 ± 13.6. The mean (±SD) Electrical Conductivity (EC) of the top 20 cm is 6.6 ± 4.0 dS/m and from 20-50 cm is 6.6 ± 5.3 dS/m. The soil texture is clayey. The development of a strong, impervious crust on the surface is characteristic of these poorer sites. The dominant vegetation species observed are gumweed (*Grindelia squarrosa*), wild barley, salt grass, and rabbitbrush (*Chrysothamnus* spp.).

On sites where the salinity, sodium concentration, and sodicity are not as limiting, the vegetation cover is 80-100%. On such sites, the mean (±SD) SAR of the top 20 cm is 5.6 ± 2.3 and from 20-50 cm is 4.8 ± 2.2. The mean (±SD) EC of the top 20 cm is 6.6 ± 3.1 dS/m and from 20-50 cm is 5.9 ± 2.9 dS/m. The soil texture ranges from loam to sandy loam. The most common vegetation species observed are brome grass and sweet clover.

Small, shallow, permanent, and vernal water bodies are common to the Type 1 ecosite. The most common vegetation species are cattails and rushes. These areas attract geese and ducks during migration, mating, and nesting seasons.

The Type 1 ecosites generally are used for pasture. The numerous small water bodies in the Old Mac area are utilized by the local retriever club to train their dogs and hold competitions.

Ecosite 2: Type 2 Spoil. Three characteristics of the Type 2 ecosite distinguish it from the other ecosites within the Spoil Pile Complex ecosection:

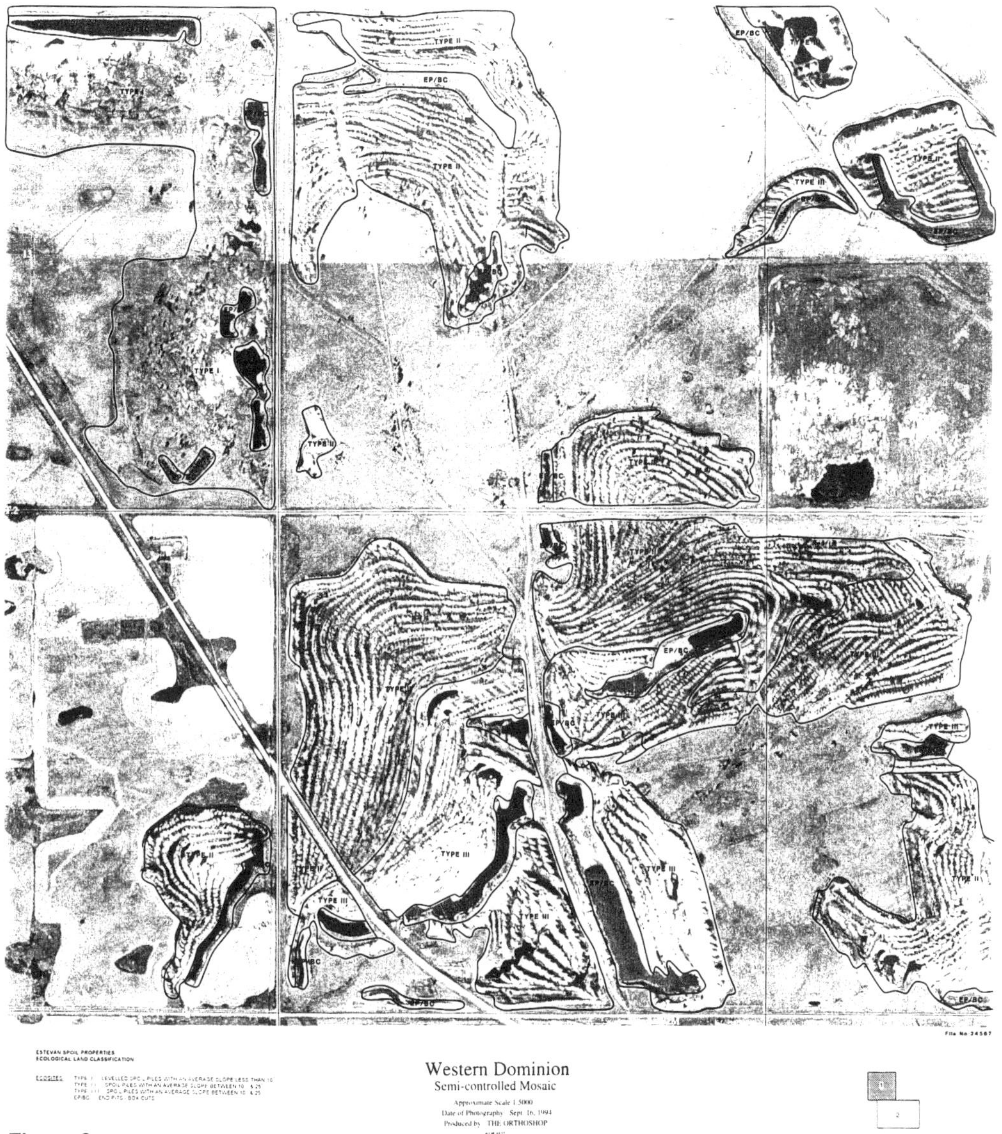

Figure 3.

- the average slope of the piles is between 10° and 25°;
- the average height is less than 7.5 m; and
- the percent vegetation cover is generally greater than 70%.

The Type 2 ecosites are primarily the result of removing overburden using shovels or small draglines. A shovel or small dragline is not capable of removing deep cuts of overburden like a larger, more modern dragline can; therefore, the slopes are gentler and shorter, minimizing surface runoff and erosion and promoting water retention.

The soil quality of spoil piles created by shovels and small draglines generally is better than that of a large dragline because a lower proportion of bedrock was incorporated into the rooting zone of the spoil pile. The lower proportion of bedrock in the rooting zone results in relatively less saline and sodic conditions. The mean (±SD) SAR of the top 20 cm is 0.7 ± 0.2 and from 20-50 cm is 1.7 ± 1.0. The mean (±SD) EC of the top 20 cm is 0.6 ± 0.1 dS/m and from 20-50 cm is 0.7 ± 0.1 dS/m. The soil texture is sandy loam to loam and the hard, impervious crust characteristic of saline, sodic sites is not observed.

The most common species of vegetation observed in this ecosite are brome grass, salt grass, speargrass, wheatgrass, snowberry, and rose.

Ecosite 3: Type 3 Spoil. Three characteristics of the Type 3 ecosite distinguish it from the other ecosites within the Spoil Pile Complex ecosection:

- the average slope of the piles is greater than 25°;
- the average height is greater than 7.5 m; and
- the percent vegetation cover rarely is greater than 50%.

The Type 3 ecosites are the result of strip-mining operations using a larger, more modern dragline. As the dragline stripped overburden to expose the deeper underlying coal seams, the resulting spoil pile is higher, steeper, susceptible to erosion from surface runoff, and poorly disposed to promote water retention.

The soil quality of spoil piles created by mining with a dragline is generally poorer than that of the other types of ecosites identified within the Spoil Pile Complex. A greater proportion of bedrock is incorporated into the rooting zone of a Type 3 spoil pile because of the dragline's ability to dig deeper into the underlying bedrock. As the proportion of bedrock increases, salinity and sodicity increases. The mean (±SD) SAR of the top 20 cm is 28.6 ± 9.0 and from 20-50 cm is 25.4 ± 10.9. The mean (±SD) EC of the top 20 cm is 8.0 ± 2.1 dS/m and from 20-50 cm is 7.1 ± 2.3 dS/m. The soil texture is clay loam to loam and a hard, impervious crust is a characteristic of these saline, sodic sites.

The influence of aspect is more apparent on the Type 3 ecosite than on the Type 2 ecosite. This is due to the greater height of the spoil pile. The major influence of aspect is on percent vegetation cover, rather than species composition. The northerly and easterly aspects of the Type 3 ecosite generally have two to three times as much vegetation cover than the southerly and westerly aspects.

The predominant species of vegetation observed include brome grass, salt grass, wheatgrass, sage, thistle, kochia (*Kochia scoparia*), and rabbitbrush. Snowberry, rose, Russian olive, and poplar (*Populus* spp.) are observed at lower densities.

Ecosite 4: Box Cuts, End Pits. The End Pit/Box Cut ecosite is the result of mining with shovels, small draglines, and larger, more modern draglines. The box cut is the first cut in the strip mining sequence and, depending on whether it is backfilled, it may or may not be visible as a pit at the end of mining. The end pit is the last cut in the strip mining sequence and generally is visible as a pit at the end of mining.

The End Pit/Box Cut ecosite is a U-shaped depression characterized by a high, steep spoil pile paralleling one side of the cut and a steep, cut bank on the other side or by two, high steep spoil piles paralleling both sides of the cut. Haul roads into the pits, termed "ramps," frequently are observed in association with the End Pit/Box Cut ecosite and, therefore, are included within this ecosite. The height and slope of the spoil and the cut of this ecosite vary depending on the type of equipment used to mine the area.

Soil characteristics vary from site to site and often reflect the type of mining equipment used. They can be sandy loam-heavy clay in texture. The SAR values range from 5.5 to 38 and EC values range from 3.9 dS/m to 8.9 dS/m. The higher SAR and EC values are associated with mining with larger draglines and lower SAR and EC values with shovel and small dragline operations.

The plant community present in the pit depression is influenced by the amount of moisture present. Two basic moisture regimes have been identified:

- Permanent Water Bodies: these depressions are charged by ground water and by surface runoff. Common vegetation species include cattail, sedge, rush, and bulrush.
- Vernal Water Bodies: these depressions are flooded during the spring by runoff and may be recharged by summer rainstorms. Common vegetation species include cattail, sedge, rush, and bulrush. These water bodies differ from permanent water bodies by the relatively lower proportion of water to vegetation.

***Wildlife habitat assessment*.**

Habitat Suitability Index. The habitat suitability of the old mined lands for mule deer and Sharp-tailed Grouse is summarized in Table 1. Each of the identified ecosites was ranked based on its HSI value.

Mule Deer. The habitat suitability ranking for mule deer for the four ecosites varied from poor to good: Type 2>Type 3>Box Cuts/End Pits>Type 1. The reduced HSI value for Type 1 was associated with shrub canopy cover less than 15% and topographic diversity less than 10% average slope. The shrub canopy cover provides an indication of the forage availability for deer utilization. A major portion of the mule deer winter diet is comprised of shrubs which are critical to overwinter survival (Cook n.d.). A shrub canopy closure greater than or equal to 30 and less than or equal to 70 is considered

Table 1. Habitat Suitability Index Values and habitat quality ranking for old mined lands

Ecosite (Habitat Type)	Mule Deer[1]		Sharp-Tailed Grouse[2]	
	HSI	Rank[3]	HSI	Rank[3]
Type 1	0.30	P	0.66	G
Type 2	0.62	G	0.77	G
Type 3	0.48	A	0.20	P
Box Cuts/End Pits	0.43	A	0.32	P
Area weighted HSI	0.46	A	0.49	A

1Winter habitat evaluation
[2]Food and reproductive habitat evaluation
[3]P=Poor (<0.4)
A=Average (0.4-0.59)
G=Good (0.6-0.79)
E=Excellent (0.8-1.0)

optimal for mule deer. Increased topographic diversity can provide mule deer with hiding cover, warm, southerly aspects, a greater diversity of vegetation types, and microsites with reduced snow depths and protection from severe winds on portions of the deer's winter habitat (Cook n.d.).

The vegetation cover and vegetation diversity of Type 2 ecosites were greater than that of the Type 3 and Box Cuts/End Pits ecosites and was the major difference which resulted in the Type 2 ecosites ranking higher than the Type 3 and Box Cuts/End Pits ecosites (good vs. average). Increased vegetation cover and diversity can provide additional forage, hiding cover, and thermal cover (Cook n.d.)

The area weighted HSI for the entire study area ranked average (0.46) for mule deer.

Sharp-tailed Grouse. The habitat suitability ranking of the four ecosites for Sharp-tailed Grouse varied from poor to good: Type 2>Type 1>Box Cuts/End Pits>Type 3 (Table 1). The low HSI values for the Box Cuts/End Pits and Type 3 ecosites appeared to be the result of poor nest/brood cover, particularly the shrub and forb component. Good nest/brood cover provides hiding cover for nesting hens and hatched eggs (Prose 1987). Maximum nest/brood cover suitability for the model occurs when the equivalent optimum area providing nest/brood cover is greater than or equal to 90% (Prose 1987). The nest/brood cover suitability continues to decrease to zero when the equivalent optimum area providing nest/brood cover is 5% (Prose 1987). For the Box Cuts/End Pits and Type 3 ecosites the equivalent optimum area providing nest/brood cover was 0.32 and 0.20 respectively.

The area weighted HSI for the entire study area ranked average (0.49) for Sharp-tailed Grouse.

Wildlife surveys. Wildlife surveys for mule deer were conducted to provide an estimate of the actual, current use of the old mined lands instead of the habitat suitability of the area. In 1993, PCL conducted a summer pellet group count to estimate the extent to which the deer used the ecosites during the summer. In 1994, PCL conducted winter aerial surveys to estimate the deer densities on the old mined lands and estimate the extent to which the deer used the ecosites for winter habitat. Although the results of the pellet group count and winter aerial survey refer to both mule deer and white-tailed deer (*Odocoileus virginianus*), more than 95% of the deer observed on the old mined land were mule deer.

Winter aerial surveys. The winter aerial surveys were conducted in January and March of 1994. The following habitat types were surveyed: floodplain (Souris River Valley), coulees and upland (surrounding native prairie and agricultural lands), and Type 1, and Type 2 and 3 combined as one habitat type. Type 2 and 3 spoils were combined into one habitat type because the inconsistent pattern and mixing of the Type 2 and Type 3 ecosites made it impractical to survey them separately.

The winter aerial survey provided estimates of:

- a relative comparison of the population density of deer on old mined lands and surrounding habitat types; and
- a relative comparison of which habitat types (ecosites) were being used as winter habitat by deer.

The estimated deer density of the mined lands supported the results of the HSI

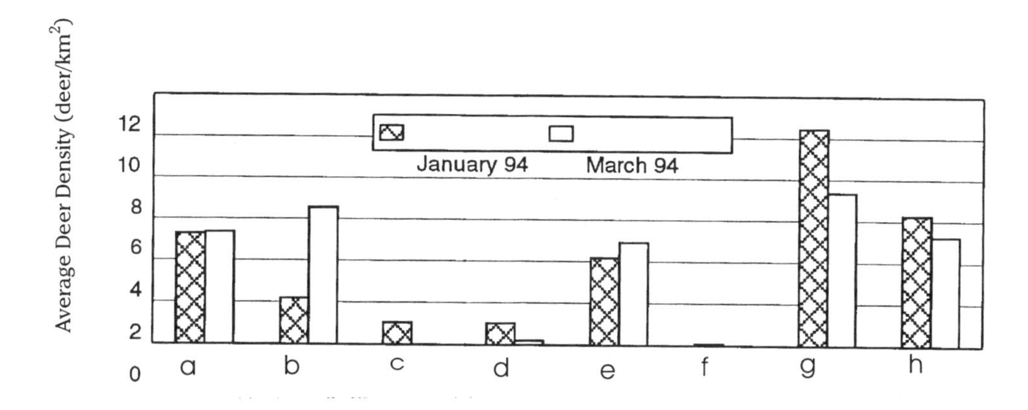

Figure 4. Estimated deer densities (deer/km^2) for old mined lands and surrounding habitats.

a. Western Dominion Spoils; b. Old Mac Spoils; c. North Costello Spoils; d. Type 1 Spoils; e. Type 2 & 3 Spoils; f. Upland; g. Coulee; h. Floodplain.

values with the deer densities of Type 2 and 3 ecosites greater than Type 1 (Figure 4). The highest deer densities were observed in the coulee and floodplain habitat types; however, during the March survey the deer densities of the Type 2 and 3 ecosites were similar to those of the floodplain (Figure 1). Most of the deer observed within the old mined lands were mule deer ($\approx 95\%$) and most of the deer observed on the floodplain were white-tailed deer ($> 95\%$). Within the coulees a mixture of mule deer and white-tailed deer was observed.

To estimate the utilization of the different ecosites, the study area was broken down into the following areas: Old Mac Mine, predominantly composed of Type 2 ecosites; North Costello, predominantly composed of Type 3 ecosites; and Western Dominion, predominantly composed of Type 2 and 3 ecosites. The deer densities observed during the surveys closely supported the results of the HSI models except in two instances. During the January survey, the highest deer densities were observed in the Western Dominion (Type 2 and 3 ecosites) and were more than twice as high as what was observed in the Old Mac area (Type 2) (Figure 4). From the model, it was expected that the Type 2 spoils would have the highest deer densities. The deer densities of the Type 2 ecosites > Type 3 > Type 1 (Figure 4).

During the March survey, the highest deer densities were observed at Old Mac Mine (Type 2), but the lowest densities were observed in North Costello (Type 3) (Figure 4). From the models, it was expected the Type 3 spoil would have higher deer densities than Type 1 ecosites. However, only one deer was observed utilizing the Type 1 ecosites and no deer were observed in North Costello (Type 3). All the deer observed in the Old Mac area (Type 1) were within a 40-hectare area on the east end of Old Mac. No signs were observed of deer utilizing the remaining 420 hectares of Old Mac.

The Western Dominion area, which is composed of the Type 2 and 3 ecosites, had the most consistent deer densities for the two surveys (Figure 4). Also, from visual observations, as many deer seemed to be using the Type 3 ecosites as there were using the Type 2 ecosites.

Summer pellet group counts. The summer pellet groups counts were useful for two reasons:

- they provided an estimate of which ecosites were being used during the summer;
- they provided an additional measure of habitat use for assessing the results of the habitat suitability index models; and
- they provided an estimate of how deer used the ecosites during the summer compared to winter.

The Box Cuts/End Pits ecosite was not analyzed separately. Any pellet groups counted in this ecosite were counted as Type 1, 2, or 3 depending on which ecosite was adjacent. The average (± SD) number of deer days use per hectare for Type 2 (4.19 ± 5.34) > Type 3 (2.10 ± 1.47) > Type 1 (0.54 ± 0.94). These results support the HSI values obtained for the three ecosites.

The same general trends were observed for summer use as during the winter aerial survey: Type 2 and 3 ecosites were being used more than Type 1.

Conclusions

The results of the ELC and wildlife studies were evaluated in terms of establishing end land use objectives for spoil property management. The habitat variables selected for the biological evaluation allowed PCL to foresee the consequences of manipulating key variables such as topographic diversity and shrub cover during reclamation and revegetation programs.

The habitat assessments were confirmed through different measures of wildlife use. They confirmed that the HSI models selected for assessing the old mined land habitats yielded essentially the same rankings of wildlife value as those determined by actual deer use measures. A finding of the habitat assessment and seasonal utilization surveys for Type 3 ecosites and the implications of managing them is that Type 3 ecosites are being utilized more by deer than is expected when there are Type 2 ecosites close by, even though the vegetation cover is minimal and the HSI ranking is on the lower end of being average (Table 1). This increased usage of the Type 3 ecosites could be attributed to increased habitat diversity which provides an additional food source in the winter during deep snow.

The ELC and wildlife habitat assessment provided an ecologically defensible justification for wildlife habitat/pasture as the best end land use for the Long-Range Spoils Management Plan. The HSI models showed that the old mined lands have, for the most part, average to good habitat potential for wildlife. Wildlife surveys showed the deer using the old mined lands as winter and summer habitat.

The soils information collected for the ELC confirmed that the agricultural potential of the Spoil Pile Complex would not be improved by full scale levelling due to the absence of topsoil. Such levelling could create large, bare, sodic, saline areas without the topographic variation and resulting microhabitats which promote revegetation and wildlife use (Kagis 1967). The topographic diversity and microhabitats created by the spoil piles provide important shelter, escape terrain, hiding cover, thermal cover, nest/brood cover, and forage for wildlife and cattle.

The ELC served to identify and establish criteria for assessing areas which PCL considered to be suitable for Release from Further Decommissioning and Reclamation obligations. Under the Mineral Industry Environmental Protection Regulations, a Release would signify that the parcel of old mined land submitted for Release has achieved an acceptable end land use.

The ELC continues to be an integral part of the management of the Estevan Properties, both in terms of land management and reclamation. The old mined lands have been used for cattle grazing for over 15 years. PCL has started to develop improved range management practices with the cooperation of their lessees. These practices include conducting annual range condition surveys, improving stocking rates based on range condition surveys, and improving grazing rotation systems. The ELC mapping has delineated the various range sites which need to be considered for conducting range condition surveys and implementing rotation systems.

The ELC has identified areas of aesthetic concerns along major traffic corridors where partial levelling and revegetation trials have been initiated. The ELC has served to provide the following baseline information and design criteria required for starting these trials and ensuring revegetation success as well as refining future reclamation techniques:

- reslope spoils to angles of maximum 25°;
- select grass/legume seed mixes which are drought and salt tolerant and provide suitable forage for cattle and wildlife; the vegetation inventory conducted for the ELC provides a list of grass and legume species to consider for reclamation;
- select a fertilizer mixture based on the soil sample analysis; and
- select trees and shrubs observed during the ELC inventory for revegetation and wildlife habitat enhancement programs.

Reclamation work is incremental and results are being evaluated on an annual basis to refine and develop cost-effective reclamation techniques relative to the land use objectives.

Prairie Coal Ltd. has submitted its Long-Range Spoil Management Plan to Saskatchewan Environment and Resource Management (June 1994). SERM endorsed the process for developing the Long-Range Spoils Management Plan and supports PCL's efforts to manage the old mined lands.

References

Abouguendia, Z.M. 1990. *A practical guide to planning for management and improvement of Saskatchewan rangeland.* Saskatchewan Research Council Report No. E-2520-1-E-90.

Cook, J.G. n.d. *Habitat suitability index models: mule deer.* U.S. Dept. Interior, Fish and Wildlife Service, Washington. 34 pp.

Demarchi, D.A. 1985. A regional wildlife ecosystem classification for British Columbia. In *Land/wildlife integration* No. 3. Compiled by H.A Stelfox and G.R. Ironside. Land Conservation Branch, Canadian Wildlife Service, and Environment Canada. pp 11-19.

Kagis, H. 1967. Third interim report on the introduction of tree cover in the Estevan spoil banks. Project S-14. Interim Report.

Prose, B.L. 1987. *Habitat suitability index models: Plains Sharp-Tailed Grouse.* U.S. Fish Wildlife Service Biological Report 82(10.142). 31 pp.

Rowe, J.S. 1979. Revised working paper on methodology/philosophy of ELC in Canada. In *Application of ecological land classification in Canada.* Edited by C.D.A. Rubec. Minister of Supply and Services Canada, Ottawa.

Strong, W.L. 1979. *Ecological land classification and evaluation - Livingstone-Porcupine.* Alberta Energy and Natural Resources, Edmonton.

Wilken, E.B. and G. Ironside. 1977. The development of ecological (biophysical) land classification in Canada. *Landscape Planning* 4: 273-282.

Restoration for Sustainability: Progress and Prognosis

by Henry T. Epp

14 Shawbrooke Court S.W.
Calgary, Alberta T2Y 3G2

Abstract. As the turn of the century nears, the role of government in environmental management is decreasing and industry is becoming increasingly self-regulated. The self-regulation is driven by consumer demands for safe products that do not contribute unnecessarily to environmental degradation. Government, representing the people, retains overall control via its legal role, but is less involved in day-to-day inspection, testing, and regulation. Environmental restoration and protection has had considerable success in heavy and resource-based industries during the last three decades, and manufacturing and transportation pollute much less than they once did over much of the world. Environmental impact assessments and long-term management plans with public input now are routinely integrated into the planning process rather than being tacked on to the ends of planning and development processes driven entirely by economic aspirations. Yet, while industry clearly has assumed some responsibility for its actions relative to the environment and the public, progress on other fronts has been less than satisfactory. A case in point is wetland conservation across the Canadian Prairies. Some progress is being made, but it is certain that most of the remaining wetlands, only about one-half the original, will disappear unless governments cease subsidizing drainage in response to lobbyists. Restoration is important, but a truly sustainable life will not be attained without maintaining overall landscape and ecosystem process integrity.

Reclamation, restoration of disturbed lands to either natural or to humanly useful states, became part of normal corporate business practice in the late 1980s and early 1990s across most of the developed world. As little as twenty years ago, however, governments reluctantly were beginning to try to force land disturbers, miners in particular, to begin reclaiming the lands and waters they had despoiled or were in the process of exploiting. This was in response to an increasingly restive public that was beginning to recognize the widespread acceleration of environmental degradation caused by human development. Public recognition included the very real public subsidy of industries that tore riches out of the land, often went elsewhere with the profits, and then left the frequently poisonous messes for cleanup via the local public purse. In this paper, the words "reclamation" and "restoration" are used synonymously.

In the 1970s, exposure as being subsidized and ruining an increasingly valued nature did not sit well with corporate executives used to proclaiming themselves as strictly part of the private sector economy and as suppliers of employment and other economic contributions to communities. Their corporate image slumped severely. Something had to be done.

First, attempts were made to cajole governments into dropping or ignoring their own legislation and to discredit environmental stakeholder groups. This failed, as the public mood had shifted permanently. At last, spurred on by fear of legal liability and a desire to improve corporate images, and, possibly, by some feelings of responsibility for their own actions, corporate management began to incorporate environmental restoration into their business plans, finally accepting this as part of the cost of doing successful business.

Environmental reclamation or restoration as the century and millennium tick to a close is on track, with observable trends occurring elsewhere as well in the world of environmental science and management. In fact, events and progress are proceeding apace and on so many fronts at once that distinct trends with long-term potential are difficult to identify. In this regard, Cortner and Moote (1994) indicate that "the emerging paradigm does appear to be based on two principles: ecosystem management and collaborative decision making."

Ecosystem management as an endeavour may be narrowed down to management for long-term resource and ecological sustainability, a goal clearly identified by the World Commission on Environment and Development (1987) and the International Union for the Conservation of Nature and Natural Resources (IUCN) (1991) after extensive worldwide public consultation. An additional trend that is accelerating as I write is a shift away from environmental regulation by governments to self-regulation by industry. At first glance this appears to be a case of "placing the fox in charge of the hen house" as environmental stakeholders are wont to say. However, a detailed look at the reality of causes and events reveals a different picture.

Consumers, especially in the developed world, are demanding that the goods they purchase be produced by corporations and processes that incorporate environmental protection and restoration into project planning as well as into implementation and decommissioning. This is manifested in the formation of the International Standards Organisation (ISO), which is busy developing consumer standards for corporations desiring to do international business (cf. Brown 1996). This, plus fear of liability at national and local levels, has driven this serious attempt at self-regulation by industry. And it is having success already. The Canadian forest industry, for example, is finding it difficult to market pulp, paper, and timber products in Europe unless it can prove forest restoration. It is busy doing this right now by preparing long-term forest management plans and environmental impact statements, and by increasing public involvement in planning and decision making.

Another and really important instance of progress in Canada is recognition of aboriginal involvement in environmental decision making and management, part of the trend toward more collaborative decision making. The need for and level of desired aboriginal involvement is exemplified by the March 8-10, 1995 Churchill River Conference held in Saskatoon, Saskatchewan. The presenters made it clear that they do not consider themselves to be stakeholders, environmental or otherwise. They insisted that they must be *part* of any decision-making process pertaining to lands and waters used by them for traditional pursuits. They consider themselves to be more than providers of input only into any decision-making process that affects their communities. Berkes (1995) identifies this need in his contribution to the conference proceedings.

Clearly, in the 1990s it behooves corporations planning to exploit resources impinging upon aboriginal and other communities to involve local people in their entire process, from planning through implementation to decommissioning and restoration. And this applies equally to mining, forestry, and all other large-scale resource use industries.

A weakness in the concept of sustainability is that it cannot be applied equally to all environmental restoration projects. For example, sustainability in mining reclamation can refer only to restoring the mined lands as near as possible to their original state to approximate ecosystem, agricultural, or community sustainability, depending on the original conditions of the lands involved. The mined ore, however, cannot be returned, so this aspect of reclamation does not address sustainability. Moreover, the mined-over lands can never be restored to their precise pre-mining situations as materials have been removed permanently, so total reclamation does not occur. Bringing in fill is not a long-term solution either, as fill materials must come from some other place which itself then is disturbed permanently. Retaining large-scale landscape integrity, including drainage patterns, may be an ideal goal of reclamation and may be possible, but local conditions are changed permanently, and this fact must be accepted if mining is to occur at all.

Restoration of wetlands and other ecosystems disturbed by water flow manipulation is easier, physically, than is reclamation of mined lands. Water flow regimes that have been altered artificially can be re-altered to restore the original flow patterns. This may not be politically or economically practical in all cases, however, even when water use patterns change, especially where people have become dependent upon the altered flow patterns. Compromises may be necessary.

In the 1990s, the concept of reclamation has broadened to become environmental restoration generally, including ecosystem restoration in which the focus is on ecological processes. In this respect, reclamation includes even the attempts at artificially regenerating populations of endangered species and then returning them to the wild. Examples of this sort of effort that come to mind include the Siberian tiger, California Condor, Peregrine Falcon, and Whooping Crane. Such efforts can meet with success, however, only following restoration of the respective ecosystems to conditions conducive to survival not only of the endangered species but of others on which they depend. At this time, the long-term survival of the endangered species mentioned here, and most others, remains moot due to human pressures on the requisite lands and waters.

Finally, it is necessary to address the fundamental question of whether ecosystem restoration can be successful. Are there examples of once destroyed ecosystems that have been reclaimed and now are functioning sustainably? Surprisingly, the answer is a resounding yes! This answer is surprising to many due to media hype, since the mass media always prefers a good story to the much duller truth. As a result, unverified hypotheses about deteriorating environments almost always receive more media coverage than do "good news" stories, and public perception of reality, including environmental problems, becomes skewed.

Easterbrook (1995) has documented dozens of cases the world over where environmental restoration has proceeded quickly and with success, in many instances simple cessation of human interference sufficing, in other instances deliberate actions being required. He demonstrates clearly that in the United States major strides have been made on the air pollution, acid rain, hazardous chemicals, and habitat protection and restoration fronts, that per capita energy consumption is much lower than twenty years ago, and that endangerment of species is being addressed successfully in many instances and has been vastly exaggerated in others.

Closer to home, all strip mines in Saskatchewan now must restore the mined-out landscapes to close approximations of their original contours and ecological conditions, be they natural or agricultural. Moreover, proponents of new mines now are required to file decommissioning plans with the provincial government, and must deposit funds for this before start-up, in case of future financial insolvency. This situation is not unusual; it is part of a general trend throughout the developed world.

In Saskatchewan's forests, as of 1992, forest harvesting companies must file comprehensive environmental impact statements with the provincial government prior to approval of long-term (twenty year) forest management plans. In these impact statements, the companies must demonstrate that they will maintain the overall ecological patch mosaic and ecosystem processes (i.e. ecosystem integrity) in perpetuity in addition to maintaining a sustainable long-term timber supply in their license areas (Epp 1995). The emphasis on ecosystem and timber supply sustainability is similar elsewhere in Canada and the rest of the developed world, but with less attention to environmental impact assessments and their consequent statements.

On the wetland reclamation front on the Canadian prairies, success still lags well behind continued wetland degradation due to drainage for agricultural and water management purposes. It is here that Ducks Unlimited and the Saskatchewan Wetland Conservation Corporation, for example, have been busy identifying key wetland sites and problem areas. They have been reconstructing some damaged wetlands and protecting others, but drainage continues apace, often spurred on by provincial governments that at once subsidize and promote it while trying to prevent it in other instances. At this time, over 50% of the prairie and parkland wetlands in Saskatchewan, the province with the greatest area of such wetlands in Canada, is gone already (Epp 1992).

Clearly, the politics of drainage need to be ironed out before wetland restoration and protection result in maintenance of at least the *status quo*. The Canadian prairies and parklands region is a living example of reasonable environmental stakeholder groups being overwhelmed by strident and much more aggressive pro-drainage lobbyists. Perhaps Easterbrook's (1995) exposure of distorted and often hyperaggressive lobbying by environmentalists in the United States was necessary, but exposure of the opposite situation is necessary too. The Canadian prairies wetland restoration effort clearly is in great need of a much firmer approach to be as effective as some of the other environmental lobbies have been in the developed world (see Fuller and Riemer, this volume). The wetlands situation is so grave that surely the truth should have some shock effect. Or have both the environmental lobbyists and the governments become too jaded to pay serious attention to this need?

The papers presented in this book focus on the scientific aspects of environmental reclamation or restoration. This concluding paper helps place the scientific focus into a wider cultural milieu which includes technology. In conclusion, clearly, ecological restoration, as we approach the turn of the century and the millennium, involves:

(1) the need for a sound scientific groundwork to establish environmental baselines and activities and materials required to re-establish these baselines to their closest possible approximation following exploitation;
(2) public and scientific input into the entire resource use process from planning to restoration;
(3) more than stakeholder involvement by affected communities and aboriginal people in the entire environmental management process;
(4) self-regulation by industry in line with increasingly environmentally conscious international consumers; and
(5) maintaining landscape integrity and ecosystem processes as fundamental to long-term environmental and resource sustainability.

References

Berkes, F. 1995. The role of co-management in conservation planning. In *The Churchill: a Canadian heritage river:* 202-208. Edited by P. Jonker. Proceedings from the Conference held on March 8-10, 1995, Saskatoon, Saskatchewan. Extension Division, University of Saskatchewan, Saskatoon.

Brown, G. 1996. ISO-14000 - it's coming - get ready! *CMA Magazine,* March 1996:6.

Cortner, H.J. and M.A. Moote. 1994. Trends and issues in land and water resources management: setting the agenda for change. *Environmental Management* 18:167-173.

Epp, H.T. 1992. Saskatchewan's endangered spaces and places: their significance and future. In *Saskatchewan's endangered spaces: an introduction:* 47-108. Edited by P. Jonker. Extension Division, University of Saskatchewan, Saskatoon.

Epp, H.T. 1995. Application of science to environmental impact assessment in boreal forest management: the Saskatchewan example. *Water, Air* and *Soil Pollution* 82:179-188.

Easterbrook, G. 1995. *A moment on the earth: the coming age of environmental optimism.* Penguin Books, New York.

International Union for the Conservation of Nature and Natural Resources (IUCN). 1991. *Caring for the earth: a strategy for sustainable living.* Gland, Switzerland.

World Commission on Environment and Development. 1987. *Our common future.* Oxford University Press, Oxford.